GW01246624

BRITISH CERAMIC PROCEEDINGS
NO. 62

CERAMICS: FROM PROCESSING TO PRODUCTION

ALSO FROM IOM COMMUNICATIONS

Ferroelectrics 2000 UK

Novel Chemistry and Processing of Ceramics

Ceramic Interfaces

Ceramic Interfaces 2

Grain Boundaries: Their Character, Characterisation and Effect on Properties

CERAMICS: FROM PROCESSING TO PRODUCTION

Proceedings of the
Annual Convention of the Ceramic Industry Division of
The Institute of Materials
Royal Agricultural College, Cirencester
10–11 April 2000

Edited by

Julie Yeomans
The University of Surrey, UK

Book B0720
First published in 2001 by
IOM Communications Ltd
1 Carlton House Terrace
London SW1Y 5DB

IOM Communications Ltd
is a whollyowned subsidiary of
The Institute of Materials

ISBN 1-86125-147-5

Typeset in the UK by
Fakenham Photosetting, Norfolk

Printed and bound at
The University Press, Cambridge

Contents

Preface

Not only was 2000 heralded by many as the start of the new Millennium, it was the Centenary Year for the British Ceramics Community. The Ceramic Industry Division of the Institute of Materials holds an Annual Convention and the 2000 event took place on the 10 and 11th of April at the Royal Agricultural College, Cirencester. The aim of the gathering was to celebrate the anniversary of the founding of the British Ceramic Society one hundred years ago and to look towards future achievements.

The Ceramic Science Committee of the Division organised a two day programme around the theme of 'From Processing to Production'. There were sixteen oral presentations and a number of posters. These spanned the whole range of issues that are important in the manufacture and subsequent use of ceramic articles, from powder production, through shaping and sintering to the use of fractography in identifying the causes of component failure.

This proceedings volume contains seven papers from the oral presentations. Between them they provide an interesting insight into the development of materials and techniques which are taking us forward into the 'second century'.

I would like to express my gratitude to the authors and reviewers who have contributed their time and expertise to the production of scientific content of this volume and to Peter Danckwerts, of IOM Communications, who has overseen the production of the physical volume.

Julie Yeomans
University of Surrey.

Preparation of Ceramic Foams from Polymeric Precursors

X. BAO, M. R. NANGREJO and M. J. EDIRISINGHE
Department of Materials, Queen Mary & Westfield College, University of London, Mile End Road, London E1 4NS, UK

ABSTRACT

This paper presents an overview of our research on the preparation of monolithic and composite non-oxide ceramic foams derived from polymeric precursors. Two new methods to product ceramic foams from polymeric precursors were developed. Firstly, synthesis of silicon carbide (SiC) foams as a product of direct pyrolysis of polysilane precursors and, secondly, preparation of SiC and silicon carbide-silicon nitride ($SiC–Si_3N_4$) composite foams using polymeric precursor solutions are described. The features of the ceramic foams produced were dependent on the compositions and the structures of the precursors. In the case of direct pyrolysis, the best SiC foams were produced using precursors with pyrolytic yields between 50–60 wt%. Ceramic foams produced by the polymeric precursor solution method have well-defined open-cell structures with the strut cross-sections free of macro-voids.

1. INTRODUCTION

There is considerable interest in engineering ceramic foams which possess a number of favourable properties such as low density, low thermal conductivity, thermal stability, high specific strength and high resistance to chemical attack and are suitable for industrial applications such as high temperature thermal insulation, hot gas particulate filters, hot metal filters, catalyst supports and cores in high temperature structural panels.[1–4]

Ceramic foams can be produced by different methods such as the polymeric sponge-ceramic slurry method[2,5–7], chemical vapour deposition of ceramics on to a porous carbon skeleton[8], sol-gel processes that develop porosity during phase transformations and chemical reactions,[9] siliciding carbon foams,[10] a gel-cast foam process which combines the foaming of ceramic suspensions and *in-situ* polymerisation,[11] a replication process where polymer is injected into a porous substrate such as sodium chloride which is removed later to produce carbon and silicon carbide foams[12–14] and co-blowing a solution containing a preceramic polymer and polyurethane precursors to produce foams.[15,16] Several workers[17–19] have also obtained ceramic foams by pyrolysing polymeric precursors, but this has not been exploited as a foaming method and no systematic studies on the relationship between the structures of the polymeric precursors and the characteristics of the ceramics synthesised are available.

The most common process for producing ceramic foams is the ceramic slurry method, i.e., impregnation of polyurethane foams with slurries containing ceramic particles and appropriate binders followed by pyrolysis and pressureless sintering at elevated temperatures. However, ceramic foams produced by this method are of low strength and fracture toughness as very thin webs with coarse-cells of ceramic and a hole in the centre of the strut cross-sections are normally left after the organic sponge is burned out, making them sensitive to stresses and limiting their structural applications.[2,20,21] The use of polymeric precursors to produce ceramic foams offer distinct advantages over the ceramic slurry method, mainly simplicity of processing and ease of control of the structure of the foam.

In the present paper, two new processes for producing SiC foams and SiC–Si_3N_4 composite foams from polymeric precursors are described. Further details of this work can be found in references 22–26. Firstly, a series of polysilanes with different substituents were synthesised and pyrolysed in an inert atmosphere to produce SiC foams directly. The relationship between the structures of the polymeric precursors and the features of the ceramics produced are also discussed. Secondly, SiC foams have been prepared by immersion of polyurethane foams in a polymer precursor solution to form pre-foams where dichloromethane was used as a solvent. After drying, the pre-foams were subjected to firing at a high temperature to obtain SiC foams. This method is also exploited further to prepare composite foams and SiC–Si_3N_4 is used as an example. The quantitative estimation of the porosity and mechanical properties of the foams prepared is being investigated at present.

2. EXPERIMENTAL DETAILS

2.1. *Synthesis of polysilanes*

The polysilanes discussed in this paper were synthesised by the alkali dechlorination of various combinations of chlorinated silane monomers in refluxing toluene/tetrahydrofuran with molten sodium as described previously.[27–29] Details of the various monomers used are summarised in Table 1.

2.2. *Preparation of SiC pre-foam*

A polyurethane (PU) sponge with open-cells in the size range 400–800 μm was used in this work. 0.8 g of polysilane was dissolved in 2000 mm^3 of dichloromethane to form a polymeric precursor solution. The PU sponge was first cut into 10 mm long cubes and these were then immersed in the precursor solution to form SiC pre-foams which were air-dried overnight at ambient temperature and subsequently pyrolysed in nitrogen.

Table 1. Monomers used in the synthesis of polysilanes

Monomers	Formula	Abbreviations
Dichloromethyphenylsilane	$(CH_3)(C_6H_5)SiCl_2$	MP
Dichloromethylsilane	$(CH_3)HSiCl_2$	MH
Dichlorophenylsilane	$(C_6H_5)HSiCl_2$	PH
Trichloromethylsilane	CH_3SiCl_3	TCM
Trichlorophenylsilane	$C_6H_5SiCl_3$	TCP
Dichloromethylvinylsilane	$(CH_3)(CH_2 = CH)SiCl_2$	MVin

2.3. *Preparation of SiC–Si_3N_4 pre-foams*

After the polysilane precursor (PS5) was dissolved in dichloromethane to prepare a solution, Si_3N_4 powder in the size range 0.1–4 μm (supplied by the AME Division of Morgan Matroc Ltd., Stourport-on-Severn, UK) was added to the solution. Typically, 0.4 g of polymer was dissolved in 2000 mm^3 of dichloromethane and then 0.1 g of Si_3N_4 powder was added to this solution. The proportion of Si_3N_4 powder in the solution was varied. The polyurethane sponge was first cut into 10 mm long cubes and these were then immersed in the polymer/Si_3N_4 powder suspension to produce SiC/Si_3N_4 pre-foams. The samples were air-dried overnight at the ambient temperature and were pyrolysed subsequently in nitrogen.

2.4. *Pyrolysis*

Pressed pellets (~8 mm diameter and ~3 mm thick) of the polymeric precursors and the pre-foams were placed in an alumina boat and heated from the ambient temperature to 900°C at 1°C min^{-1} in a tube furnace in the presence of flowing nitrogen gas (flow rate ~2.5 × 10^5 mm^3 min^{-1}) followed by soaking at this temperature for 2 hours. Then the furnace was switched off and allowed to cool to the ambient temperature. Some of the pyrolysed samples were heated further to different final temperatures (1100°C to 1700°C) at 2°C min^{-1} followed by soaking at this temperature for 2 hours and, then cooling to the ambient temperature at 2°C min^{-1} in the tube furnace in the presence of flowing nitrogen gas (~2.5 × 10^5 mm^3 min^{-1}).

2.5. *Characterisation*

The molecular weights of the polysilanes were determined by gel permeation chromatography (GPC) carried out at RAPRA Technology Ltd., Shrewsbury, UK. GPC studies were calibrated using polystyrene standards with chloroform as the eluent. The flow rate used was 1000 mm^3 min^{-1}. Fourier transform-infrared (FT-IR) spectra of as-synthesised polymer samples were obtained using a Nicolet 710 spectrometer. 1 mg of each polymer was ground and mixed with 150 mg of dried KBr powder and pressed into a pellet. Spectra were obtained in the range of 4000–400 cm^{-1} with a resolution of 4 cm^{-1}.

The pyrolytic yield from each polymer, PU foam, Si_3N_4 powder, SiC and SiC–Si_3N_4 pre-foams was measured by thermogravimetry. Samples were heated from the ambient temperature to 900°C in flowing nitrogen (flow rate ~500 mm^3 min^{-1}) at 10°C min^{-1} in a Hi-Res Modulated TGA 2950 thermogravimetric analyser to determine the pyrolytic yield.

The macro-appearance of the discs after pyrolysis was photographed. The microstructures of the pyrolysed products were investigated using a Cambridge S360 scanning electron microscope (SEM). Samples studied using the SEM were coated with gold.

X-ray diffraction (XRD) was carried out on the as received Si_3N_4 powder, the polysilanes and polysilane/Si_3N_4 composites after pyrolysis to various temperatures (≥1100°C) in nitrogen. Samples for X-ray diffractometry were ground using an agate pestle and mortar. A modified Philips X-ray diffractometer with filtered Cu$K\alpha$ radiation of wavelength 0.15418 nm was used (a graphite monochronometer removes $K\beta$ radiation). The voltage and current settings of the diffractometer were 40 kV and 30 mA, respectively. The scan range was from 10° to 90° with a step size of 0.05° and a scan speed of 0.025°s^{-1}.

Table 2. Details of monomer(s) used in the synthesis, polymer yield, molecular weight and pyrolytic yield of each polymer prepared

Polymer	Monomer(s) used (mol %)	Polymer yield (wt %)	$\bar{M}_w$	$\bar{M}_n$	$\bar{M}_w/\bar{M}_n$	Pyrolytic yield (wt %)
PS1	MP = 100	43	7680	1520	5.1	23.6
PS2	MP/TCM = 70/30	60	4270	1840	2.3	30.4
PS3	MP/MVin = 70/30	37	6970	1670	4.2	42.1
PS4	MP/MVin/TCM = 60/20/20	28	3570	1560	2.3	51.0
PS5	MP/MVin/MH = 60/20/20	38	12000	2460	4.9	53.2
PS6	MP/MVin/TCM = 55/15/30	52	4240	1510	2.8	56.8
PS7	MP/MVin/TCM = 55/30/15	45	5240	1600	3.3	59.8
PS8	PH/MVin/TCP = 60/20/20	70	7700	1580	4.9	66.4
PS9	PH/MVin/TCM = 60/20/20	32	8700	2580	3.4	67.9

3. RESULTS AND DISCUSSION

3.1. Polymer synthesis and characterisation

The details of the polymeric precursors synthesised are given in Table 2.

As an example, a typical FT-IR spectrum of a ter-polysilane (PS7) is shown in Figure 1. It exhibits characteristic C–H stretching between 3100 and 2700 cm^{-1}. The peaks at 3050 and 3067 cm^{-1} represent C–H stretching in the phenyl group. Methyl group stretching is observed at 2956 and 2894 cm^{-1}. Additional peaks at 1406 and 1248 cm^{-1} are characteristic of the asymmetric and symmetric bending modes, respectively, of CH_3 bonded to silicon. Three small peaks at 1949, 1887 and 1815 cm^{-1} are attributed to the phenyl–Si vibration. The peaks at 697 cm^{-1} for Si–C stretching and 464 $^{-1}$ for Si–Si is typical for these polysilanes.[30–32] A characteristic peak is clearly present at 1589 cm^{-1} and corresponds to the presence of the vinyl group.[33, 34] The appearance of a Si–H stretching peak around 2100 cm^{-1} implies that PS7 contains some hydrosilane groups formed during polymerisation.[29]

3.2. Conversion to ceramic.

The thermogravimetric traces of the Si_3N_4 powder, PU foam, PS5 and the PS5–Si_3N_4 pre-foams are shown in Figure 2. The thermal decomposition of the PU foams starts at about 200°C and is almost fully pyrolysed at 500°C, while there is no weight loss occurring in the Si_3N_4 powder at <900°C. In the case of the polymeric precursor PS5, the pyrolysis process takes place in three consecutive stages. In the first stage (up to 350°C), a very slow rate of weight loss of less than 2% occurs and this corresponds to the loss of low molecular weight

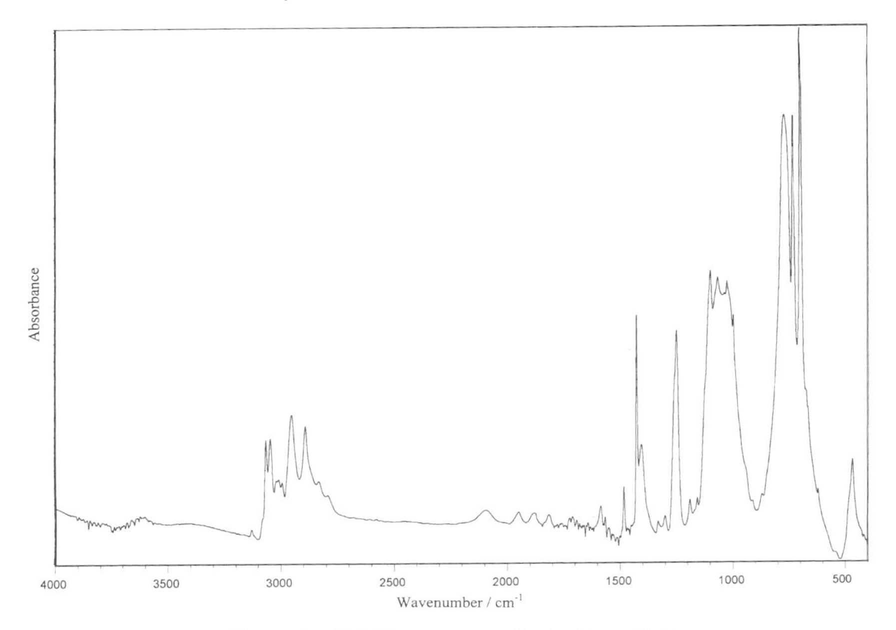

Figure 1. FT-IR spectrum of polysilane PS7.

oligomers in the polymer. In the second stage (350–600°C), a major weight loss takes place due to the decomposition and re-arrangement of the polymers and a variety of gaseous species including hydrogen (H_2), hydrocarbons (CH_4, C_2H_6, C_6H_6 etc.) and silanes ($(CH_3)_4Si$, $(CH_3)_3SiH$, $(CH_3)_2SiH_2$, etc.) are evolved.[33–37] In the third and final stage of pyrolysis (above 600°C, a further weight loss of about 2% occurs as the sample is heated to 900°C. The pre-foam of polysilane/Si_2N_4 powder shows a similar thermogravimetric trace to PS5. The major weight loss occurs between 350°C and 600°C.

It is apparent (Table 2) that the pyrolytic yield is very dependent on the composition of the polymers. Co-polymers (PS2 and PS3) and ter-polymers (PS4–PS9) give better pyrolytic yields, compared with the homopolymer (PS1). This can be attributed to the cross-linked structures formed during the pyrolysis due to the thermal cross-linking capability of hydrosilane (Si–H) and vinyl (CH_2 = CH) groups in the polymeric precursors and the branched structures generated during polymerisation due to the addition of trichloromethylsilane or trichlorophenylsilane monomers.[33–39] Such cross-link formation will allow the material to maintain its original morphology during pyrolysis because the melting of the polymeric precursors is hindered[25,40] and this is critical in preventing the collapse of the ceramic foams during pyrolysis.

XRD results showed that in the case of all the polysilanes pyrolysed (~900°C) an amorphous ceramic residue was produced and further heating is necessary to obtain crystallisation of SiC.[28,29] The XRD patterns of Si_3N_4 powder and the PS–Si_3N_4 foams heated to ≥1100°C are shown in Figure 3. The results suggest that SiC formed by the pyrolysis of the polysilanes remained amorphous after heating in nitrogen up to 1100°C and then gradually crystallised as the temperature was increased above 1300°C. Three additional characteristic peaks at $2\theta = 36°, 61°$, and 72° are observed in the crystallised material and these

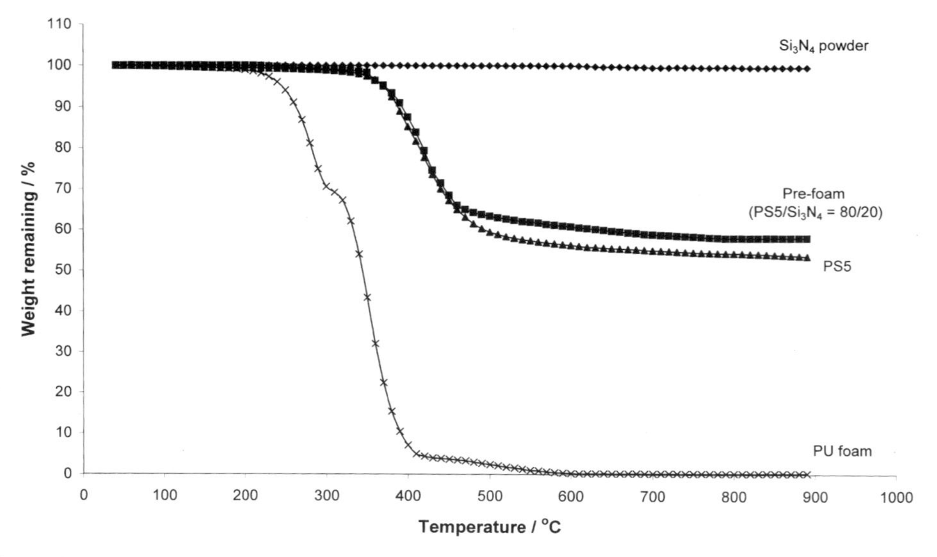

Figure 2. Thermogravimetric traces of PU, Si_3N_4 powder, polysilane precursor PS5 and pre-foam with PS5/Si_3N_4 = 80/20 wt%.

correspond to the (111), (220), and (311) planes of β–SiC, respectively,[41,42], indicating the formation of SiC–Si_3N_4 composite *in-situ*. However, as the temperature increases to 1600°C, the XRD pattern shows prominent peaks mainly at 2θ = 36°, 61°, and 72° corresponding to β–SiC and caused by residual carbon left on pyrolysis reacting with the Si_3N_4 according to the reaction given below.[43]

$$Si_3N_4 + 3C \xrightarrow{1450^\circ C} 3SiC + 2N_2 \tag{1}$$

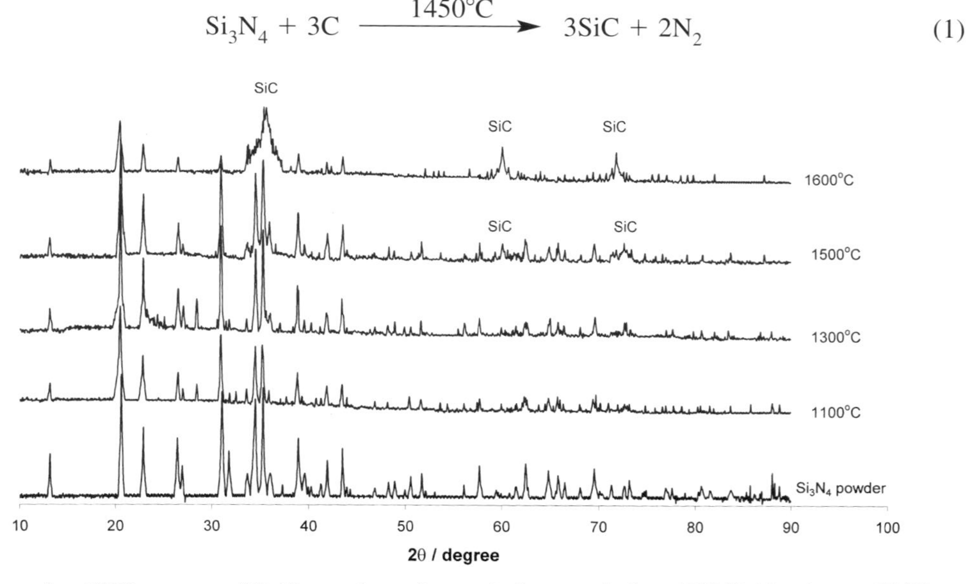

Figure 3. XRD patterns of Si_3N_4 powder and ceramic foam made from PS5/Si_3N_4 mixture (80/20 wt%) after heating to different temperatures in the range of 1100–1600°C.

3.3. *Formation of SiC foam*

3.3.1. *SiC foam produced by direct pyrolysis*

It is generally believed that pore formation in the ceramics derived from polymeric precursors is due to dissociation reactions which occur during pyrolysis of the polymers where small molecules such as hydrogen, hydrocarbons and silanes are formed and initially exist in the dissolved state.[44] When the preceramic polymer becomes supersaturated with the volatiles, nucleation and formation of gas bubbles occur. The process of conversion from a polymer to ceramic restrains and eventually arrests the growth of such bubbles which ultimately become pores in the final ceramic product.[44,45]

The formation of micropores or macropores will depend largely on the chemistry and structure of the polymer and the pyrolysis conditions such as the heating rate. As discussed in Section 3.2, the evolution of small, mainly cyclic fragments occurs from about 200°C. Evolution of gas which included hydrogen, hydrocarbons and silanes maximises in the temperature range 350 to 600°C due to the cleavage of chemical bonds, decomposition and re-arrangement of the polymer. Depending on the structures of the polymers, crosslinking also takes place in this range of temperatures.[33–39] It is believed that the relatively weak intrachain Si–Si bond undergoes cleavage about ca. 400°C leading to silyl radicals. These radicals then presumably induce cross-linking via olefin coupling involving the vinyl groups in the polymers or Si–Si/C–H bond re-arrangement processes. The presence of Si–H bonds in the polymer chain can also enhance cross-linking during pyrolysis, thus resulting in an improved ceramic yield. Gas evolution continues to ~900°C and amorphous SiC is formed.

The gaseous species formed during pyrolysis are removed by diffusion through a network of pore channels.[46] However, as the polymer crosslinks and densifies, the evolution of pyrolysis gases by diffusion faces more resistance. It can be estimated that hundreds of volumes of gases are evolved per volume of polymer pyrolysed. The rate of gas evolution at a heating rate of $2°C\,min^{-1}$ is too fast for diffusion through the solid phase to be an important route for gas loss.[47] Therefore, the bulk ceramic products derived from polymeric precursors generally consist of large pores. The higher the volume of gases evolved, the larger the size of pores formed. Only in thin fibres or films does conversion to ceramic occur without macroscopic foaming.[46,48]

It was confirmed by our study that the pore sizes in the SiC foams formed increased with the decrease of the pyrolytic yield of the polymeric precursor. The precursors with lower pyrolytic yields (e.g. PS1 and PS2 in Table 2) formed SiC foams with pore size ranging from 700–1800 μm and bloating was observed as they lost their original disc shape during pyrolysis. On the other hand, the precursors with higher pyrolytic yields (e.g. PS8 and PS9 in Table 2) produced a more densified SiC with micro-pores in the size range 5–20 μm. These ceramics maintained their original shape very well with some shrinkage. The best ceramic foams were obtained by pyrolysing polymeric precursors giving intermediate pyrolytic yields. These produced a partially open pore-structure but with a more uniform pore distribution and a smaller pore size in the range 400–800 μm (Figure 4a). These foams also retained their original shape reasonably well during pyrolysis (Figure 4b).

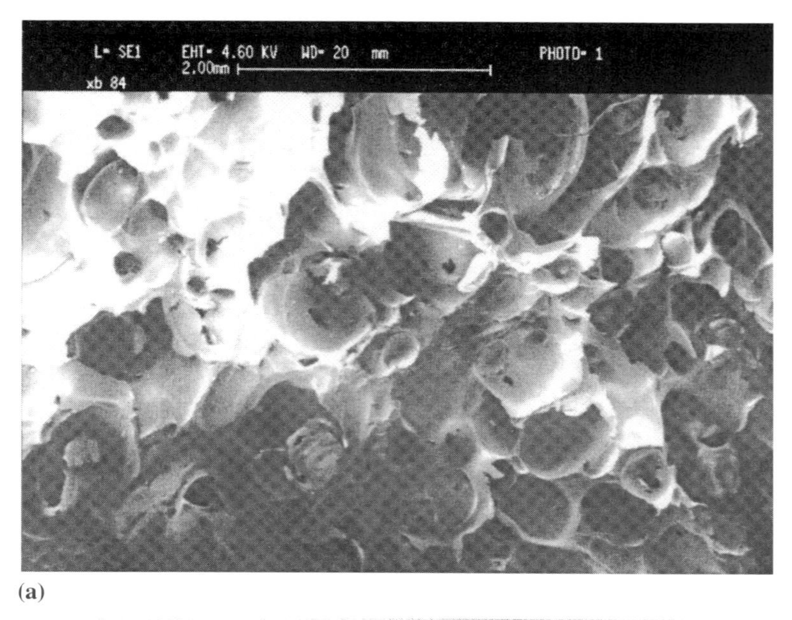

(a)

(b)

Figure 4. (a) Scanning electron micrograph of cross-section and (b) macro-appearance of pyrolysed polymeric precursor PS5.

3.3.2. SiC foam produced using precursor solutions

The structures of the polyurethane foam and pre-foam produced from PS5 are shown in Figure 5. The PU foam shows a reticular structure with open-cells in the size range 500–800 μm (Figure 5a). The pre-foam displays a similar foam structure, but some cell

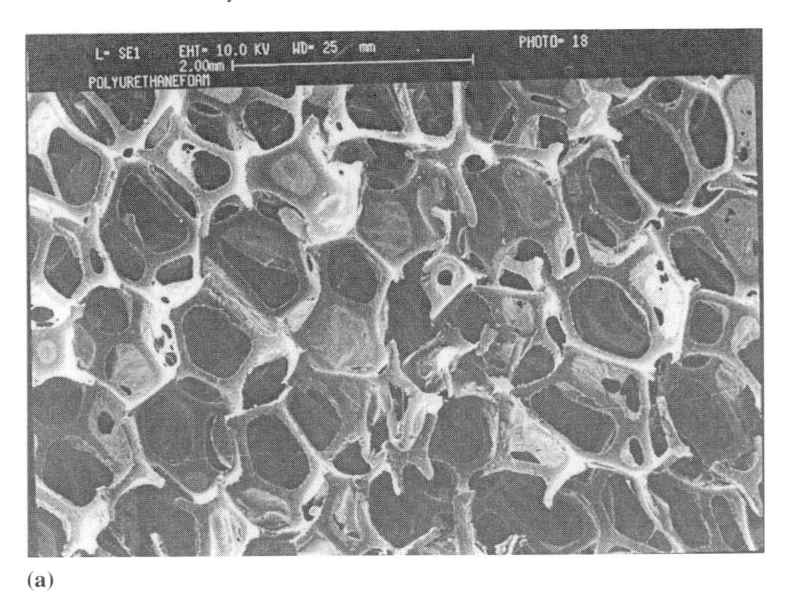

(a)

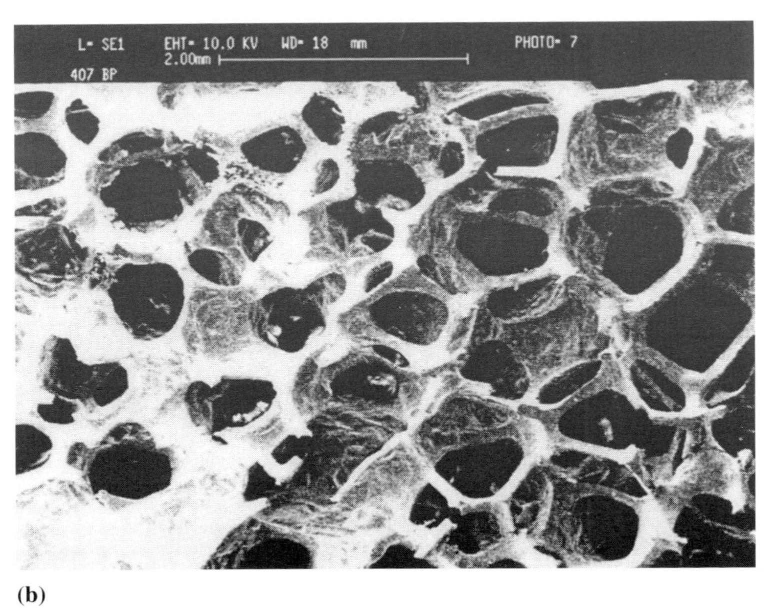

(b)

Figure 5. Scanning electron micrographs of (a) PU and (b) PS5 pre-foam.

windows are covered by a thin membrane of the polysilanes (Figure 5b). The final SiC foam produced by pyrolysis of the pre-foam consists of a three-dimensional array of struts and a well-defined open-cell structure with cell sizes between 400 μm and 800 μm. The cell windows vary in size from 200 μm to 500 μm (Figure 6a). As discussed above, the PU foam starts to thermally decompose from about 200°C and is almost fully pyrolysed at 500°C

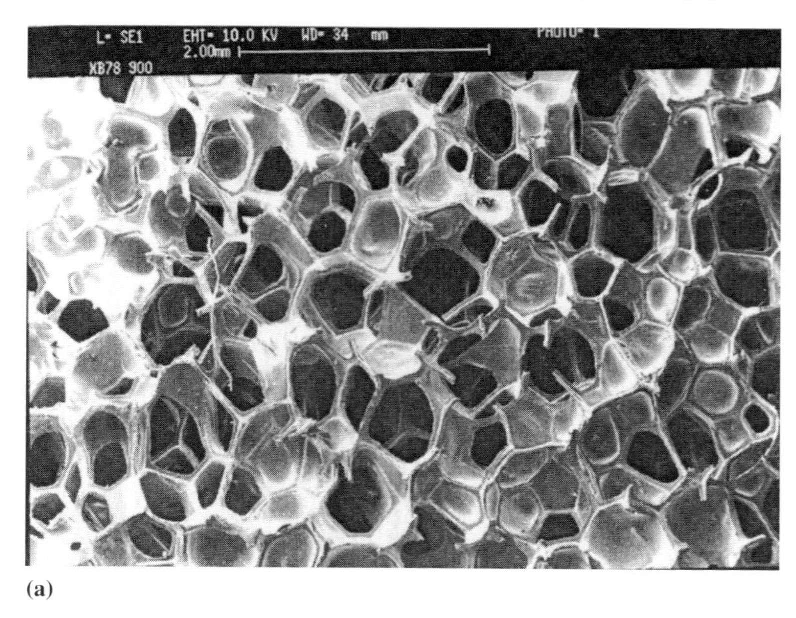

(a)

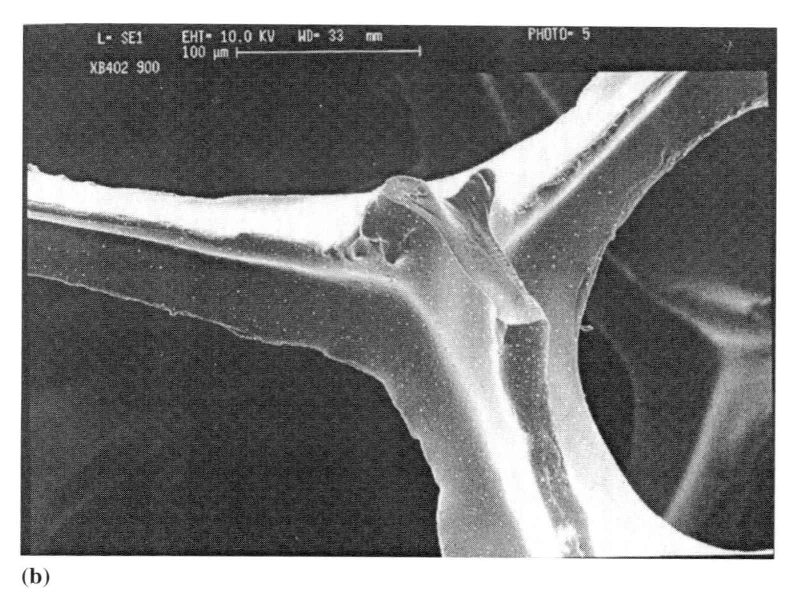

(b)

Figure 6. Scanning electron micrographs showing (a) the open-cell structure and (b) the solid strut structure of the SiC foam (after pyrolysis at 900°C) produced from polymeric precursor solution method using PS5.

(Figure 3). Hence the foam is self-supporting and retains its shape during the later stage of pyrolysis from polymer to ceramic in the temperature range 500°C–700°C. This indicates that the polymeric precursors form a homogeneous and continuous structure within the polyurethane template. The strut of the SiC foam (Figure 6b) does not show any surface cracks usually present in foams prepared from ceramic slurries.[49] Such cracks are likely to be caused mainly by the non-uniform coating of the polymeric foam by the ceramic slurry.[49,50] The struts are key building units of the foam structure and defects in them lower the strength of the foams considerably.[7]

It is noteworthy that there is no hole at the centre of the strut. In contrast, ceramic foams made by ceramic slurry coating method can contain such defects after pyrolysis[7,20,21,49,50] and will result in inferior mechanical properties. The elimination of the cross-section hole by our processing method is probably due to the better penetration of the precursor solution into the PU web structure during coating and the inward mobility of the polymeric precursor during pyrolysis. The polysilane precursor used does not fully cross-link until about 600°C and therefore such movement is possible during heating.[23,28,29]

The SiC foam was heat-treated further to 1100°C and 1700°C in nitrogen to investigate the effect of higher temperatures on the structure. Micrographs (Figure 7) show that there are no obvious changes in the morphology of the ceramic foams after heat treatment at a higher temperature. However, some pores are closed as more shrinkage takes place with the increase of temperature, due to the loss of free carbon and crystallisation of the SiC.[29]

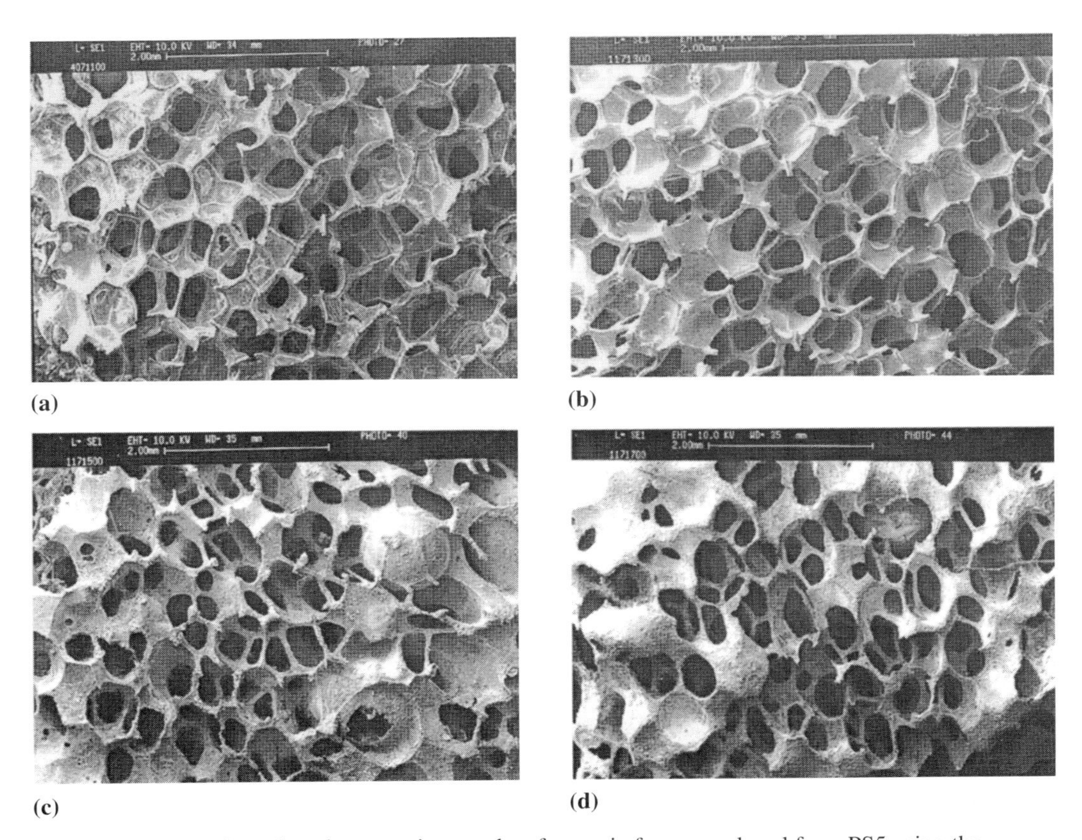

Figure 7. Scanning electron micrographs of ceramic foams produced from PS5 using the precursor solution method after heat treatment at (a) 1100°C, (b) 1300°C, (c) 1500°C and (d) 1700°C.

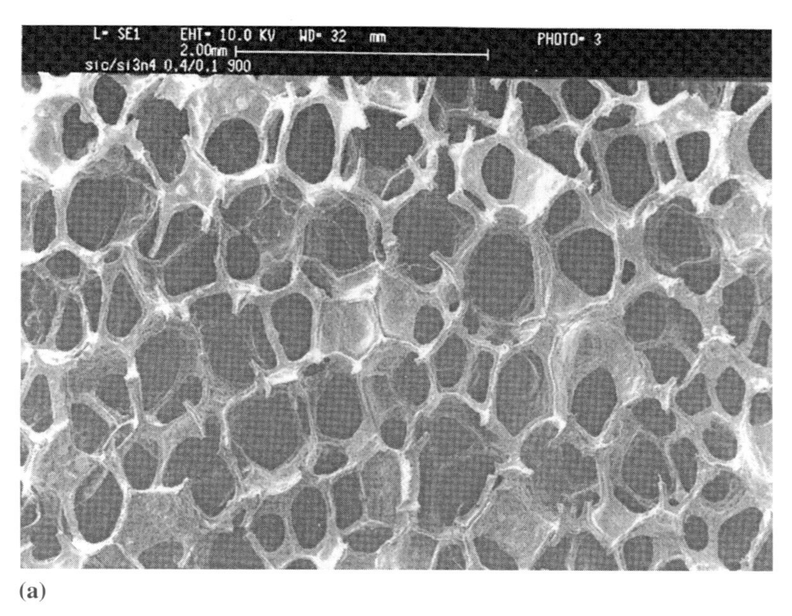

(a)

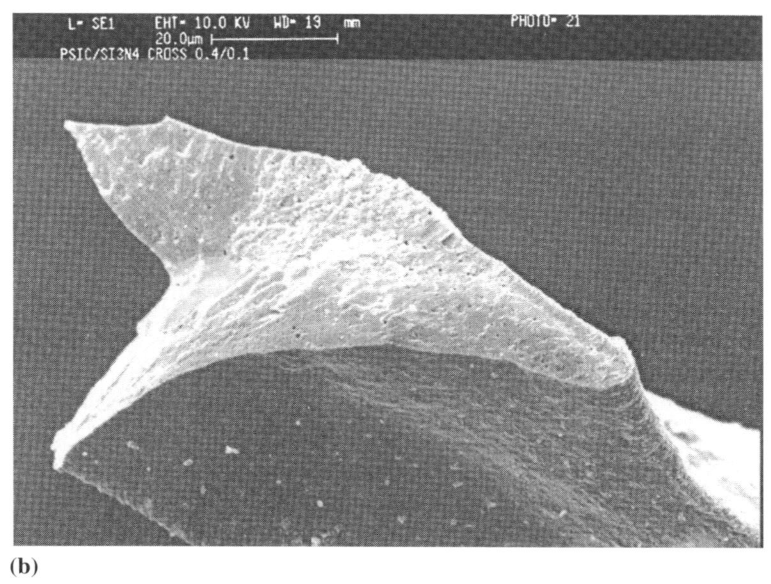

(b)

Figure 8. Scanning electron micrographs showing (a) the open-cell structure and (b) the solid strut structure of the ceramic foam produced from PS5/Si_3N_4 mixture (80/20 wt%) after pyrolysis at 900°C.

3.3.3. SiC–Si_3N_4 foams produced using precursor solutions

The SEM micrograph (Figure 8) indicates that the SiC–Si_3N_4 composite foams show similar structures to the SiC foams produced by this method (Figure 6), i.e. a well-defined reticular structure (Figure 8a) with macro-void free struts (Figure 8b).

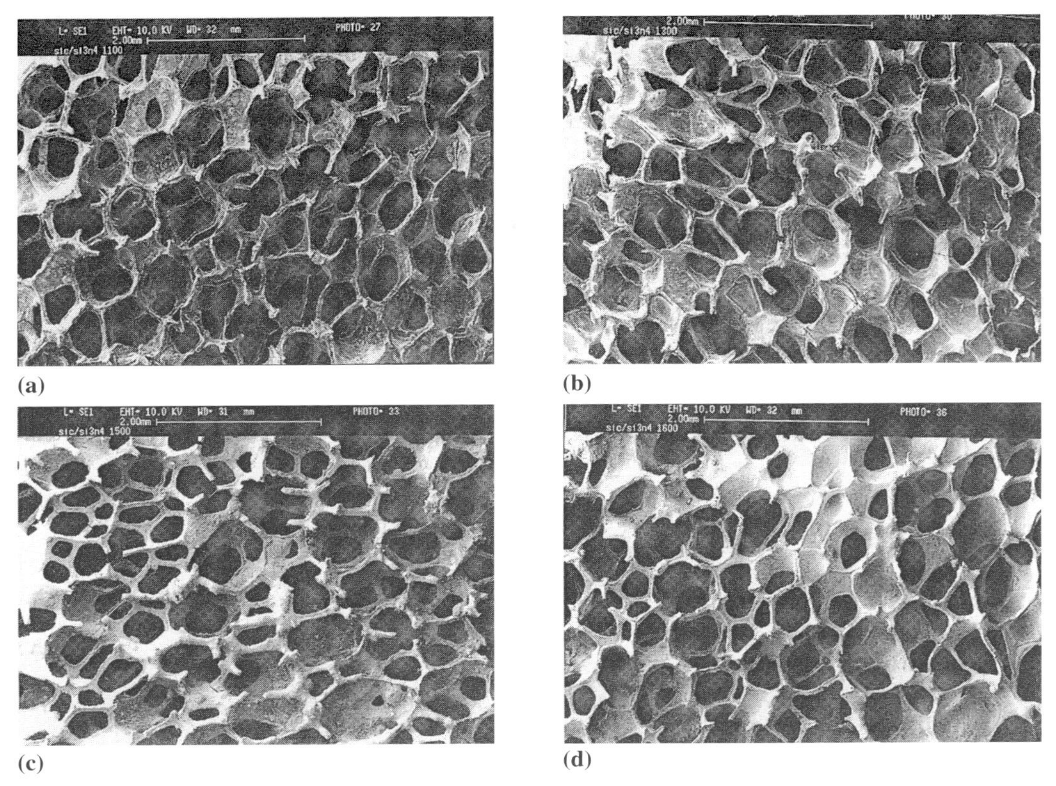

Figure 9. Scanning electron micrographs of ceramic foams produced from PS5/Si_3N_4 mixture (80/20 wt%) after heat treatment at (a) 1100°C, (b) 1300°C, (c) 1500°C and (d) 1600°C.

Micrographs of these foams after heat treatment at different temperatures (1100 to 1600°C) are shown in Figure 9. The results indicate that there are obvious changes in the morphology of the ceramic foams after heat treatment at a higher temperature. However, more shrinkage takes place with the increase of the heating temperature, particularly $>$1500°C, probably due to the loss of free carbon, crystallisation of the ceramics and the release of nitrogen gas because of the reaction between silicon nitride and free carbon as shown in equation (1).[29,43]

The influence of the percentage of the Si_3N_4 particles in the pre-ceramic polymer solution on the structure and shrinkage of the ceramic foams was also investigated. As shown in Figures 10, with increases in the Si_3N_4 particles, the overall structure of ceramic foams remains similar but the volume shrinkage of the foams is reduced (Table 3). However, at a high level of Si_3N_4 in the composite (e.g. see arrows in Figure 8d) cracking was observed in the foam structure.

4. CONCLUSIONS

Two new methods have been developed to produce ceramic foams from polymeric precursors. Firstly, silicon carbide foams have been prepared successfully by direct pyrolysis of

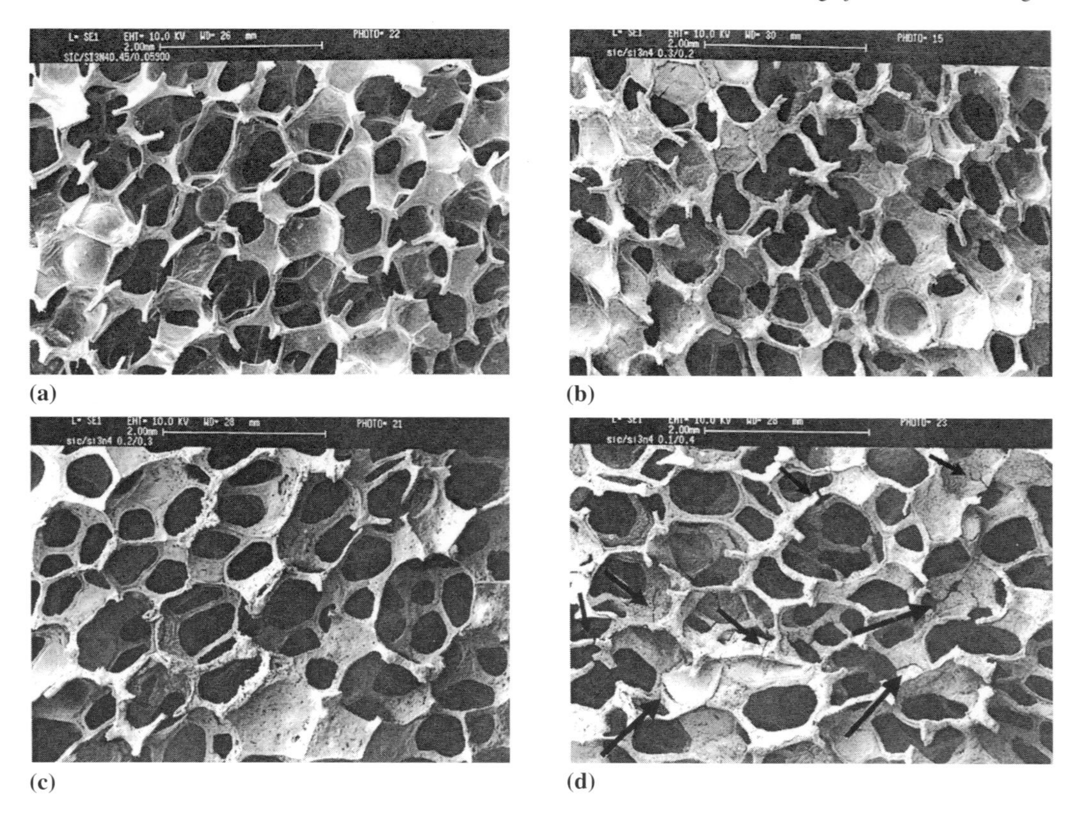

Figure 10. Scanning electron micrographs of pyrolysed (900°C) ceramic foams produced with different PS5/Si_3N_4 suspensions (wt%), (a) 90/10, (b) 60/40, (c) 40/60 and (d) 20/80. Cracks in (d) are shown using arrows.

Table 3. Linear[a] and volume[b] shrinkage on pyrolysis (900°C) of pre-foams with different proportions of Si_3N_4 in the composite. X, Y and Z refer to length, width and height, respectively, of the cubic foam.

Si_3N_4 powder in PS/Si_3N_4 (wt %)	ΔX%	ΔY%	ΔZ%	ΔV%
0	32	32	34	69
10	31	31	32	67
20	27	26	27	60
40	20	20	21	50
60	8	7	8	21
80	5	5	6	15

[a]$\Delta L\% = 100 \times (L_{\text{pre-foam}} - L_{\text{ceramic foam}})/L_{\text{pre-foam}}$
[b]$\Delta V\% = 100 \times (V_{\text{pre-foam}} - V_{\text{ceramic foam}})/V_{\text{pre-foam}}$

polymeric precursors. It was found that features of the ceramic foams produced were dependent on the compositions and the structures of the precursors pyrolysed. The best SiC foams were produced using polysilane precursors with pyrolytic yields between 50–60 wt%. These foams have partially open pores in the range of 400–800 μm.

Secondly, SiC and SiC–Si_3N_4 foams have also been produced by immersing polyurethane foam in polymeric precursor solutions or polymeric precursor/Si_3N_4 powder suspensions to form pre-foams followed by heating in nitrogen. The polymeric precursors were thus converted to SiC. The SiC and SiC–Si_3N_4 foams produced by this method showed well-defined open-cell structures and void-free struts. With the increase of temperature from 900°C to 1600°C, there were no obvious structural changes but shrinkage increased. The shrinkage of the foams reduced as the amount of Si_3N_4 powder in the composite increased but some cracks in the foam structure were observed at a high percentage of Si_3N4.

ACKNOWLEDGEMENTS

The authors wish to thank the Government of Pakistan for partial support of this work via a PhD scholarship to Mr. Nangrejo.

REFERENCES

1. R. W. Rice: 'Porosity of Ceramics', Marcel Dekker Inc. 1998.
2. J. Saggio-Woyansky, C. E. Scott and W. P. Minnear: *Amer. Ceram. Soc. Bull.*, 1992, **71**, 1674.
3. E. J. A. E. Williams and J. R. G. Evans: *J. Mater. Sci.*, 1996, **31**, 559.
4. P. Sepulveda and J. G. P. Binner: *J. Euro. Ceram. Soc.*, 1999, **19**, 2059.
5. F. F. Lange and K. T. Miller: *Adv. Ceram. Mater.*, 1987, **2**, 827.
6. R. Brezny and D. J. Green: *J. Amer. Ceram. Soc.*, 1989, **72**, 1145.
7. S. B. Bhaduri: *Adv. Performance Mater.*, 1994, **1**, 205.
8. A. J. Sherman, R. H. Tuffias and R. B Kaplan: *Amer. Ceram. Soc. Bull.*, 1990, **70**, 1025.
9. M. Wu and G. L. Messing: *J. Amer. Ceram. Soc.*, 1990, **73**, 3497.
10. Y. Aoki and B. McEnaney: *Brit. Ceram. Trans.*, 1995, **94**, 133.
11. P. Sepulveda: *Amer. Ceram. Soc. Bull.*, 1997, **76**, 61.
12. R. W. Pekala and R. W. Hopper: *J. Mater. Sci.*, 1987, **22**, 1840.
13. J. D. Lemay, R. W. Hopper, L. W. Hrubesh and R. W. Pekala: *Mater. Res. Soc. Bull.*, 1990, **15**, 19.
14. T. J. Fitzgerald and A. Mortensen: *J. Mater. Sci.*, 1995, **30**, 1025.
15. P. Colombo, M. Griffoni and M. Modesti: *J. Sol-Gel Sci. Tech.*, 1998, **13**, 195.
16. P. Colombo and M. Modesti: *J Amer. Ceram. Soc.*, 1999, **82**, 573.
17. B. E. Walker, Jr, R. W. Rice, P. F. Becher, B. A. Bender and W. S. Coblenz: *Amer. Ceram. Soc. Ceram. Bull.*, 1983, **62**, 916–923.
18. M. Seibold and P. Greil: *J. Euro. Ceram. Soc.*, 1993, **13**, 105.
19. R. Reidel: 'Advanced Ceramics from Inorganic Polymers' in *Materials Science and Technology, A Comprehensive Treatment, 17B,* 1996, 1.
20. D. A. Hirschfeld, T. K. Li and D. M. Liu: *Key Engineering Mater.*, 1996, **115**, 65.
21. J-M. Tulliani, L. Montanaro, T. J. Bell and M. V. Swain: *J. Amer. Ceram. Soc.*, 1999, **82**, 961.
22. X. Bao, M. R. Nangrejo and M. J. Edirisinghe: *J. Mater. Sci.*, 1999, **34**, 2495.
23. M. R. Nangrejo, X. Bao and M. J. Edirisinghe: *J. Mater. Sci. Lett.*, 2000, **19**, 787.
24. X. Bao, M. R. Nangrejo and M. J. Edirisinghe: *J. Mater. Sci.*, in press.

25. M. R. Nangrejo, X. Bao and M. J. Edirisinghe: *J. Euro. Ceram. Soc.*, in press.
26. M. R. Nangrejo, X. Bao and M. J. Edirisinghe: *Int. J. Inorg. Mater.*, in press.
27. X. Zhang and R. West: *J. Polym. Sci. Chem.*, 1984, **22**, 159.
28. X. Bao, M. J. Edirisinghe, G. F. Fernando and M. J. Folkes: *J. Euro. Ceram. Soc.*, 1998, **18**, 915.
29. X. Bao, M. J. Edirisinghe, G. F. Fernando and M. J. Folkes: *Brit. Ceram. Trans.*, 1998, **97**, 253.
30. J. P. Wesson and T. C. Williams: *J. Polym. Sci. Polym. Chem.*, 1979, **17**, 2833.
31. J. P. Wesson and T. C. Williams: *J. Polym. Sci. Polym. Chem.*, 1980, **18**, 959.
32. J. P. Wesson and T. C. Williams: *J. Polym. Sci. Polym. Chem.*, 1981, **19**, 65.
33. D. J. Carlsson, J. D. Cooney, S. Gauthier and D. J. Worsfold: *J. Amer. Ceram. Soc.*, 1990, **73**, 237.
34. F. I. Hurwitz, T. A. Kacik, X-Y. Bu, J. Masnovi, P. J. Heimann and K. Beyene: *J. Mater. Sci.*, 1995, **30**, 3130.
35. E. Bouillon, F. Langlais, R. Pailler, R. Naslain, F. Gruege, J. C. Sarthou, A. Delpuech, C. Laffon, P. Lagarde, M. Monthioux and A. Oberlin: *J. Mater. Sci.*, 1991, **26**, 1333.
36. R. M. Laine and F. Babonneau: *Chem. Mater.*, 1995, **5**, 260.
37. Q. Liu, H-J. Wu, R. Lewis, C. E. Maciel and L. V. Interrante: *Chem. Mater.*, 1999, **11**, 2038.
38. H-J Wu and L. V. Interrante: *Chem. Mater.*, 1989, **1**, 564.
39. R. J. P. Corriu, D. Leclercq, P. H. Mutin, J. M. Planeix and A. Vioux: *Organometallics*, 1993, **12**, 454.
40. Y. Hasegawa, M. Iimura and S. Yajima: *J. Mater. Sci.*, 1980, **15**, 720.
41. W. R. Schmidt, L. V. Interrant, R. H. Doremus, T. K. Trout, P. S. Marchetti and G. E. Maciel: *Chem. Mater.*, 1991, **3**, 257.
42. Y-T. Shieh and S. P. Sawan: *J. Appl. Polym. Sci.*, 1995, **58**, 2013.
43. Y. D. Blum, K. B. Schwartz and R. M. Laine: *J. Mater. Sci.*, 1989, **24**, 1707.
44. M. Blander and J. L. Katz: *AIChE J.*, 1975, **21**, 833.
45. H. Yao, S. Kovenklioglu and D. M. Kalyon: *Chem. Eng. Comm.*, 1990, **96**, 155.
46. J. Lipowitz, J. A. Rabe, L. K. Frevel and R. L. Miller: *J. Mater. Sci.*, 1990, **25**, 2118.
47. J. Lipowitz, H. A. Freeman, R. T. Chen and E. R. Prack: *Adv. Ceram. Mater.*, 1987, **2**, 121.
48. P. P. Loh, X. Bao, M. R. Nangrejo and M. J. Edirisinghe: *J. Mater. Sci. Lett.*, 2000, **19**, 587.
49. V. R. Vedula, D. J. Green and J. R. Hellman: *J. Amer. Ceram. Soc.*, 1999, **82**, 649.
50. D. D. Brown and D. J. Green: *J. Amer. Ceram. Soc.*, 1994, **77**, 1467.

The Use of Fractography as an Aid to Advanced Technical Ceramic Material Development and Component Design

R. MORRELL
National Physical Laboratory, Teddington, Middlesex, TW11 0LW, UK
M. MURRAY
†Morgan Matroc Ltd, Bioceramics Unit, Rugby, Warwickshire, CV21 3QR, UK

ABSTRACT
Fractography has long been applied to metallic and glass components as a tool to explain why failures occur, but it is only in the last two decades that it has been successfully applied to ceramic materials. In part this may be due to the experimental difficulties experienced with many of the coarser-grain or porous products. However, with the development of strong, fine-grained materials in which the fracture origins can be of many different types, understanding which features limit the potential performance of a material under development, or of a component in service, becomes particularly important. In recent years, standardised practices for fractogaphic evaluation have been produced, there has been an international round robin test, and a fracture toughness method reliant on fractography has been developed. This paper reviews the principles of fractography applied to advanced technical ceramic materials and components. A case study of the behaviour of ceramic femoral heads during laboratory testing is discussed in which the detailed design of the component is shown to be the performance limiting factor rather than the materials processing.

1. INTRODUCTION

Fractography is a well-established, but poorly understood and variably employed, technique for evaluating the causes of failure in a material. Requiring an element of skill and experience, it relies on visual observation, with or without microscopic aids, of fracture surfaces, and then the interpretation of the evidence obtained. It was first used extensively for studying failures in metals during the early part of the 20th century, and this is reflected in the dictionary definitions of the term, even within the last decade. It has also been used very successfully for glass materials; indeed it is in glass fractures that the finest detail of the dynamic fracture process are plainly visible. The amorphous structure of glass means that a fracture proceeds unimpeded by the irregularities found in crystalline materials, such as grain boundaries, second phases and preferred cleavage planes.

The microstructural features of many types of traditional ceramics and refractories often provide a fracture surface with such roughness that the subtleties typically seen on glass fracture surfaces are rendered invisible. In perhaps the first textbook on ceramic manufacture, principally of clay-based bodies, from the late nineteenth century, Bourry[1] advises:

> Observation of the structure or homogeneity should consist of the examination of a fracture, either by naked eye or by a magnifying glass. It will be advisable to note:- (a) the appearance of the fracture, whether granulated or rough, or with a conchoidal surface. (b)

the sizes of the grains, marked according to the classification adopted for natural stones, whether the grains are fine, medium or coarse, and uniform or of different sizes. (c) the homogeneity by observing whether the mass is entirely, moderately or slightly homogeneous, whether there are any planes of cleavage or scaling, and whether these are numerous or pronounced.

This actually represents an admirable listing of the types of features the fractographer would observe in addition to the markings attributable to the dynamics of the fracture process. However, only with finer-grained materials (those with a grain size of about 10 μm or less) can the dynamic features help the fractographer readily home in on the fracture origin and hence clarify the causes of failure.

Thus it is only the development of fine-grained advanced technical ceramics within the last two decades that fractography has been developed systematically for ceramics (e.g. Refs 2–5), to the extent that it can be now be considered as a adjunct to material development, component design and failure analysis.

This paper reviews the recent progress that has been made towards consolidating the science of fractography on ceramic materials and improving awareness of techniques and capabilities, and gives a detailed example from the biomedical field of how fractography can be used to improve product reliability.

2. THE FRACTURE PROCESS

Fracture in a brittle material without coarse microstructure commences from a site where the stress concentration is highest, i.e. usually at an appropriate stress-raiser, such as a microstructural discontinuity or a small crack. The elastic energy stored in the material and in the system loading the material then accelerates a crack from the origin, a process which absorbs energy in terms of new surface energy, generation of elastic waves and kinetic energy of fragments. When the speed of the crack is low, it tends to give a smooth fracture surface, but as the crack accelerates towards the speed of sound in the material, it tends to get rougher and rougher. This behaviour is particularly noticeable in glass, and gives rise to distinct zones surrounding the origin, as in Fig. 1. The smooth area is the 'mirror', the slightly rough area is the 'mist', and the rough area is the 'hackle'. This behaviour with increasing velocity is said to result from the driving stress concentration becoming angled to the plane of the crack, tending to divert it (see Refs 6 and 7, recently reviewed in Ref. 8). With a high-energy fracture, this can lead to forking of the fracture plane (*e.g.* Fig. 2), a process which can occur several times. When the stored energy is low, the tendency for this to happen is reduced, and fewer larger fragments are likely to be produced.

When the microstructure has a significant influence on the local fracture planes, only the coarser fracture features may be displayed. The fractographer's task is then to recognise them against the microstructural background.

In either case, the pattern of fragmentation of the material or component is related to the stress distribution applied at the time of fracture initiation, and during any time immediately afterwards. Thus in the case of a uniform uniaxial stress, simple planar cracks are likely to be seen, but this is seldom the practical situation. In a flexural strength test-piece, for

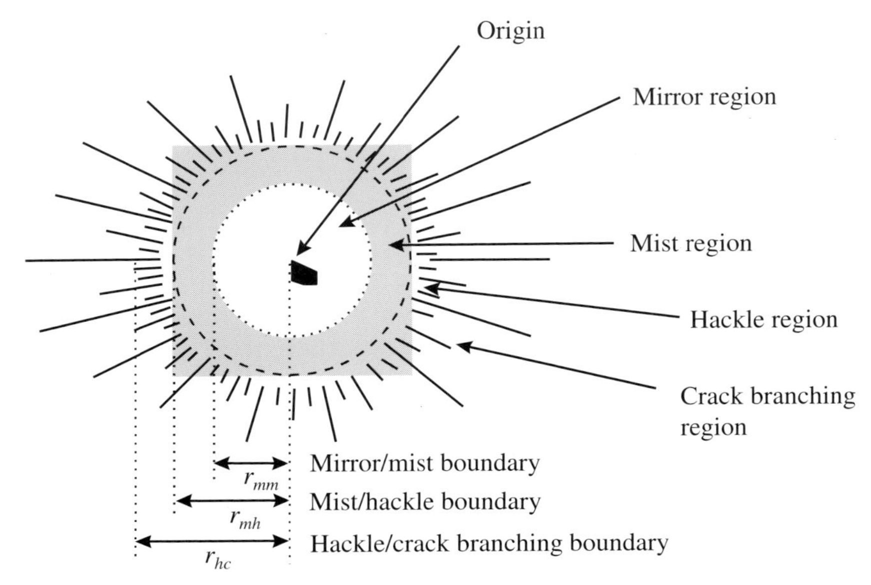

Figure 1. Schematic diagram of the development of fracture surface features on a high-strength material failing in tension from a discrete origin, together with measurements of fracture feature radius.

example, the tensile stress drops towards the middle of the test-piece, and then becomes compressive. The crack initially accelerates, but on entering the compression zone it slows down and tries to turn away, giving the so-called 'compression curl' (Fig. 2). This is a tell-tale sign of compressive stress on a fracture surface, and helps to identify the direction of stressing at the point of failure.

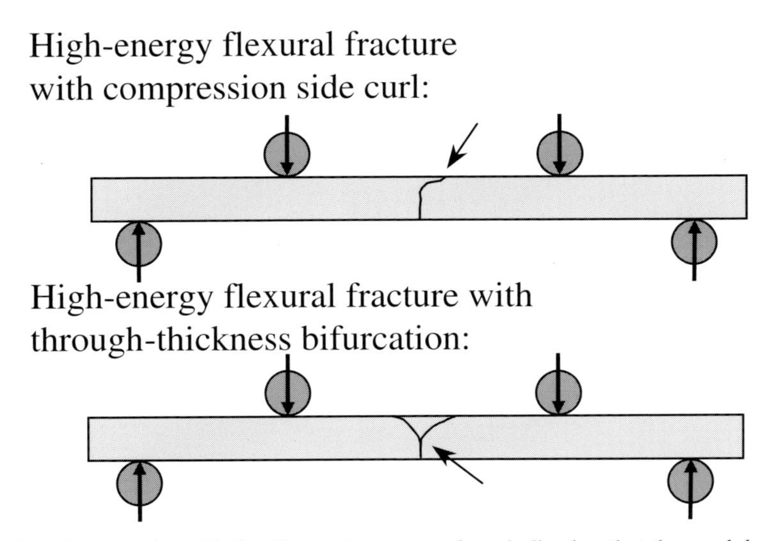

Figure 2. Compression-side 'curl' on a fracture surface, indicating that the crack has been slowed and deviated by compressive stresses.

3. THE FRACTOGRAPHER'S TASK

The fractographer has to start with the broken fragments and work in the reverse direction to the original fracture process. There are a number of distinct steps which can be summarised as follows:

1. If necessary, the fragments should be gently cleaned and inspected, and then the jigsaw puzzle is assembled. This has varying degrees of difficulty depending on the number of fragments.
2. The pattern of cracks is studied, looking for evidence of fractures which bifurcated from a common origin (e.g. Fig. 3) and distinguishing these from secondary breaks, for example as fragments impact on others or on their surroundings after the initial propagation. Other than fractures from component edges, often there is a degree of symmetry in the crack pattern which can provide an immediate focus toward the position of the origin.

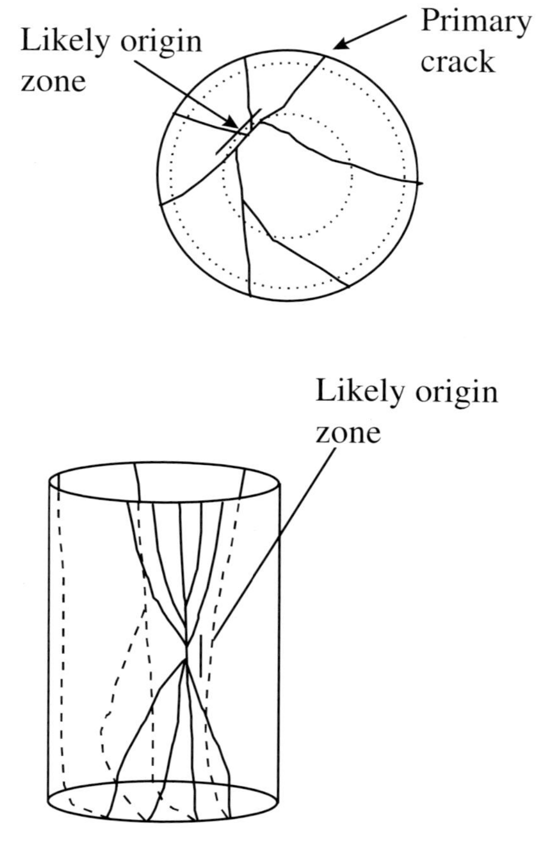

Figure 3. Typical crack patterns in (upper) a ring-on-ring biaxial fracture test on a disc test-piece, and (lower) a tube burst test. Note that the patterns contain both a primary fracture plane and a sequence of bifurcations and secondary fractures.

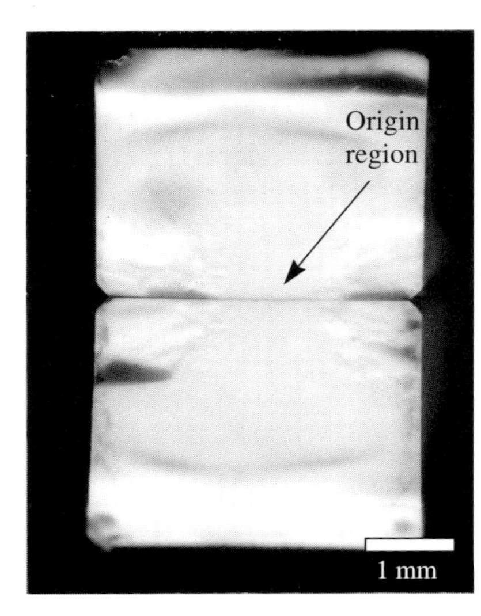

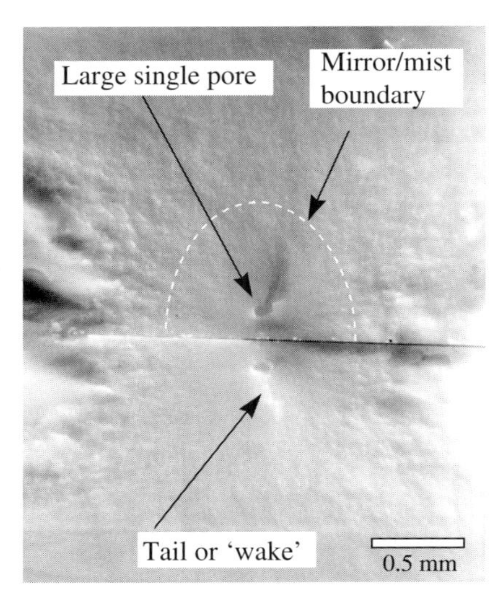

Figure 4. Matched fracture surfaces of a flexural strength test bar of a high-strength alumina ceramic showing (left) a distinct mirror region, and (right) the origin as a discrete pore with a 'wake' feature. The mirror shape is elliptical because the tensile stress declines quickly through the thickness of the test bar.

3. The actual fracture surfaces in the region of the origin can then be studied, looking for tell-tale markings which permit the position of the origin to be determined. This can often be done with the naked eye, but is easier under a low-magnification binocular microscope (Fig. 4). Features to look for, in sequence, are:
   - a relatively smooth area between hackled areas;
   - fracture markings within the smooth area which tend to radiate from a localised region;
   - the focus of the fracture markings, particularly discrete inhomogeneities.

   Of course, life may not be simple, particularly if a large area of the fracture surface is smooth, or if the fracture origin is a pre-existing crack or non-localised machining flaw.
4. Assuming clear markings are present, the origin can be examined at higher optical magnification for evidence of its nature, particularly whether it is clearly within the bulk of the material or at the surface, and whether it is highly localised as a discrete defect, or non-localised such as a pre-existing crack or machining flaw.

Often this may be as far as it is necessary to take the process if there is a simple, optically visible explanation, such as a crack or a pore. However, the optical microscope has neither the depth of focus nor the resolving power nor the chemical analytical capability of the scanning electron microscope for establishing the true nature of many types of origin, should this need to be indentified. For example, in material development, knowing whether poor mechanical performance is due to the presence of localised impurity concentrations, inhomogeneity of phase distribution or agglomerates may be crucial, and in such cases the SEM provides the tools for probing the origin more closely (*e.g.* as in Fig. 5c).

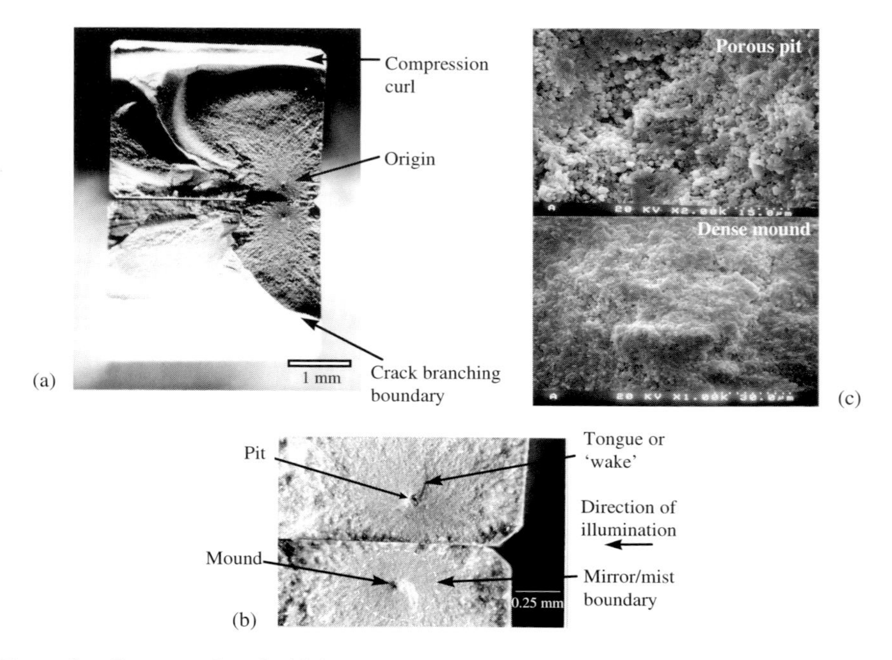

Figure 5. Fractography of a high-strength silicon nitride showing (a) matching halves of the fracture surface, (b) at higher magnification with a clear internal origin, (c) SEM identification of the origin as an agglomerate accompanied by a porous region.

Of particular value for evaluating fractures in component design, evaluation and use is the quantification of the fracture surface features, particularly the sizes of obvious origins or the fracture 'mirrors'. This process can help to identify the stress in the component at the time of fracture, or conversely, if the stress is known, what size of fracture-initiating origin or mirror to expect. There are two basic fracture mechanics equations:[9]

for the fracture mirror of radius r:

$$\sigma_f = \frac{A}{\sqrt{r}}$$

for the size of the origin, c:

$$\sigma_f = \frac{K_{Ic}}{Y\sqrt{c}}$$

where σ_f is the stress at the instant of fracture initiation, A is the so-called 'mirror constant', K_{Ic} is the critical stress intensity factor, and Y is an origin shape parameter. The similarity of these two equations also implies that there is a relationship between origin size and mirror size, which can be quite useful for correlating the respective observations.

For many materials, the mirror constant A is known to within about 30% from evaluations of flexural strength data coupled with mirror size measurements (see Annex in ASTM C1322[9]). Thus, measurement of the mirror radius allows the stress at fracture to be roughly estimated. If there is no boundary to the mirror observable within the fracture surface, the stress was too small for the crack to reach the velocity required to initiate hackle, so the origin must be large. Similar calculations of stress can be made from the size of discrete origins, if K_{Ic} is known. Conversely, if the stress is known, the origin size can be estimated, which is particularly useful if the origin is not obvious, *e.g.* a machining flaw. Thus, presented with adequate evidence, it is often possible to make a full diagnosis of the fracture circumstances, although caution is advised that fracture mechanics can provide only rough estimates because of material uncertainties and the effects of origin shape.

Of course the above overall analysis process may not provide categorical answers, particularly if:

- the coarseness of the grains obliterates the sought-for fracture surface features;
- the material is inhomogeneous to the extent that the inhomogeneity controls the fracture path, *e.g.* veins of pores originating from poorly compacted spray-dried granules;
- crucial fragments are damaged subsequent to initiation of fracture by impact on hard surfaces (Fig. 6), or are missing; collecting all the pieces and keeping them clean and separate from each other is highly desirable but often overlooked;

Figure 6. Strength test bar of a silicon nitride which has suffered impact damage on the test jig resulting in chipping of the edge containing the origin; only the small area on the lower left is original fracture surface.

- surfaces are contaminated during or after fracture; one of the worst problems for SEM analysis is the presence of mounting clays or glues used to 'reassemble' parts for initial examination; they readily contaminate the critical parts, are difficult to remove and can be readily misinterpreted as the feature of importance.

4. GUIDING FRACTOGRAPHERS

Fractography is akin to playing detective, and good detectives rely heavily on experience. While the principles behind fractography seem simple, there is no substitute for hands-on experience for quick and reliable recognition of the evidence. However, to encourage serious consideration and use of fractography, various efforts have been made worldwide. A series of fractography of ceramics and glasses conferences in the USA and their associated proceedings[2–6] have brought amateurs and professionals together to advance the science. The ASTM guidance document mentioned above (ASTM C1322[9]) has been developed as a public domain procedure which can be used as appropriate by those less experienced in the science. The document provides a classification of origin type, and includes a number of examples of how fractures originate in a variety of high-strength advanced technical ceramics, with an emphasis on flexural strength test-pieces. In support of the development of the standard, an international round robin was initiated under the auspices of VAMAS* Technical Working Area 3. This provided voluntary participants with a series of fractured test-pieces to evaluate using the prototype standard. In most cases, a limited description of the material type, its general properties and the circumstances of fracture (*e.g.* nominal strength) were given, and the participants were asked to identify the origins and dimension them using microscopy and guided by the prototype standard. The outcomes of the exercise are fully reported elsewhere,[10] and were used to modify the standard, particularly to clarify the procedures. A number of participants completely misinterpreted some of the fractures and did not use all the information they were given, indicating a definite need to improve skills and experience. A CEN standard is also being developed in Europe with a similar purpose, and this attempts to clarify further the series of steps that should be used by providing a flow diagram. The standard has gathered examples of fractures from many laboratories across Europe, in order to illustrate the different appearances that fracture surfaces and fracture origins can take.

A test procedure requiring the use of fractography is now standardised. Within ASTM, provisional standard PS 70[11] contains three methods for measuring fracture toughness, one of which is the surface-crack-in-flexure method. In this method a small pre-crack is introduced by indentation and the damage zone is then removed, leaving a pseudo-elliptical surface crack. After fracture of the test-piece, the crack size needs to be measured and the site of initiation of propagation identified. High-magnification SEM images are usually required in order to distinguish the pre-crack boundary and whether any subcritical crack growth has occurred.

NPL has recently produced a Measurement Good Practice Guide to fractography of

*Versailles Agreement on Advanced Materials and Standards, the steering committee of which has representatives from Canada, France, Germany, Italy, Japan, UK, USA, plus the European Commission acting on behalf of other European Community members.

brittle materials,[12] particularly ceramics and hardmetals, and gives many further examples, including some on ceramic components as case histories.

5. USE OF FRACTOGRAPHY TO OPTIMISE DESIGN

Strength test bars represent perhaps the most straightforward cases for fractography. Not only are the fractures usually made under controlled conditions, but the fracture stress is known and the stress distribution is (or should be) well defined. When it comes to components, this is often not the case, especially with in-service failures.

Because ceramic materials are brittle, they cannot tolerate stress concentrations in the same way as many metal alloys or tough polymers, and therefore design of ceramic components and their method of installation has to be done in a way which minimises the stresses applied. This process is often undertaken incorrectly, even if design tools such as finite element analysis are used, and this leads to poor performance. The question often arises as to whether the material was faulty, the design was faulty, or the usage was incorrect. Although the forces applied to the component are usually unknown, and there may be a very large number of fragments, with some missing and others contaminated with foreign material, fractography can help to decide which cause is most likely. In summary, the basic way in which fractography is used is as follows:

1. The general fracture pattern is symptomatic of the shape of the object, the manner in which it was stressed to failure and the site of the origin. Use the fracture pattern, particularly any symmetry, to identify the origin position
2. Examine the origin and if possible, record its nature, its size and the mirror size. When possible, use the fracture mechanical approach to estimate the stress at fracture. Decide whether the stress level was a reasonable expectation for good quality material, or unexpectedly low. If it is unexpectedly low, identify whether the origin type was intrinsic to the material, i.e. a normal part of the material microstructure, or extrinsic, caused by the way in which the material has been machined, handled or used. If it is an intrinsic origin failing at low stress, the material quality may well be suspect.
3. Using the likely stress distribution generated by the application of normal external forces, identify whether the origin is in a region of expected high or low stress. If the origin is in a nominally low stress region, but is of extrinsic type, fracture may have resulted from localised mechanical contact, impact or handling damage, such as scratching or chipping. If the origin is in a highly stressed region and shows a high fracture stress, the design may be at fault by permitting such high stresses to be applied.

In this way, the best directions for finding a solution can be quickly identified.

6. DESIGN CASE STUDY – CERAMIC FEMORAL HEADS

In order to satisfy demands for a degree of flexibility over the relative positioning of a ceramic femoral head onto an implanted metallic femoral stem to suit the patient, orthopaedic manufacturers vary the internal geometry of the cone-shaped location of the head onto the

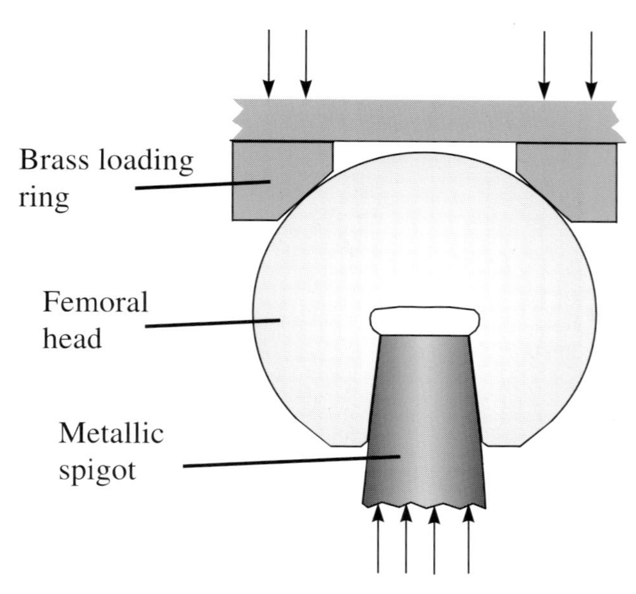

Figure 7. Schematic of the ISO 7206-5[13] and ISO 7206-6[14] mechanical test arrangement for ceramic femoral heads, for ultimate compression strength and fatigue respectively.

stem (Fig. 7). Normally there is a small angular mismatch between the conical surfaces of head and stem such that contact first occurs at the inner end of the cone surface, and as load is applied through the patient's weight, the stem deforms a little to provide a good locking contact region. This region determines the stress distribution within the head, which is basically in permanent hoop tension, peaking at the bore surface. Changes in the cone geometry and contact length affect the stress distribution in a marked manner. Although finite element analysis can provide an indication of stress level and the location of any stress concentrations, it is a difficult modelling situation with assumptions concerning frictional effects between head and stem, elastic/plastic conditions in the stem, and perfection of the geometry and cone angle mismatch. Consequently it is necessary to test mechanically the head/stem combination in order to meet health authority performance requirements (Fig. 7a). For example, for type approval by the US Food and Drugs Administration, the minimum strength of a batch of five heads loaded monotonically into axial compression onto stems must be greater than 46 kN, and a batch of three heads must survive 10^7 fatigue cycles loaded in compression at a peak force of 14 kN, after which the residual ultimate compression strength must be greater than 20 kN.
In testing a variety of designs of Y-TZP ceramic heads for use on both titanium and cobalt chrome alloy stems, variations in performance were found, some of which did not meet the target performance level in all examples in a batch (Fig. 8). It was unclear whether this was a result of changes in the manufacturing procedure with modified geometry, or whether the stress distribution was unfavourable. A possible route to identifying the source of the problem and thus improve the reliability of the designs was via fractography.

As mentioned above, a femoral head axially loaded onto a conical stem is subject to hoop

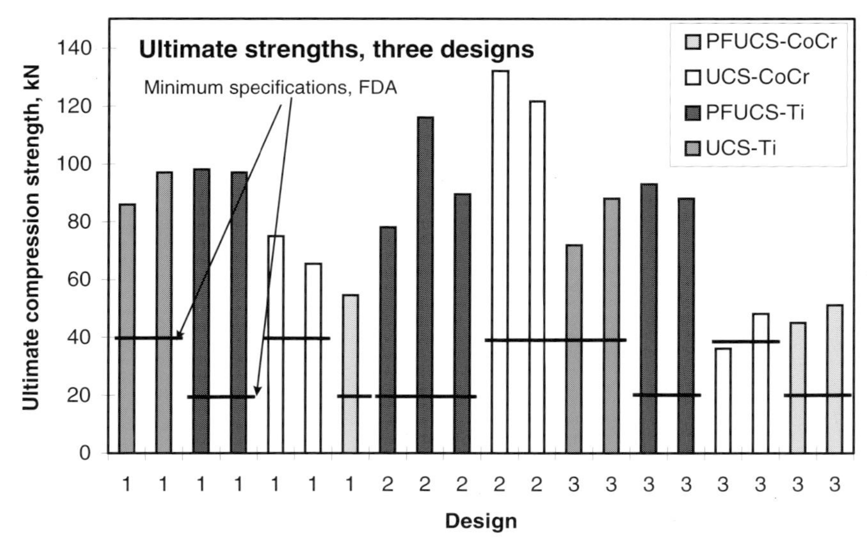

Figure 8. Ultimate compression strength (UCS) of three designs of femoral head tested in monotonic loading before and after fatigue testing at 14 kN for 10^7 cycles (post-fatigue UCS), using two different spigot materials, titanium alloy and cobalt–chrome alloy.

tension in the bore region. At the force levels required for the tests, contact between head and stem is substantial, often over the entire available area, so the position of fracture origins need not be consistent, but they are usually easy to find from the fracture pattern. When a head is deliberately fractured in an axial test, the initiating crack normally runs axially from the origin into the dome of the head (Fig. 7b), but may bifurcate depending on the fracture stress. The external forces applied to the head during the test (normally through a loading ring in accordance with ISO 7206-5, -6 (Refs. 13 and 14)) cause the fracture to slow down, deviate and bifurcate in the dome region. Then, because of the wedging effect of the stem, these cracks run back down the other side of the head, usually non-axially, to cause final failure into a number of fragments, typically 8 or 10 in total. Usually the initiating crack is planar, but the subsequent cracks show evidence of compression zone curl, allowing the initiating crack to be quickly identified. The origin can then be identified on this initiating crack surface as the focus of the smooth region between hackled regions.

In the particular case of Y-TZP, it was found that because of the high strength and toughness of the material, the mirror size is small and poorly defined (e.g. Fig. 9). In fact it appears that the roughness of the surface progressively increases on moving from the origin to the regions of strong bifurcating hackle, and does not show distinct boundaries which are so often characteristic in other materials, such as fine-grained alumina (cf. Fig. 4).

In order to tackle this problem, a set of flexural strength test bars of known strengths was examined. A visual criterion was developed for identifying the size of the mirror as the radius of the circle with a minimum level of contrast in photographs taken with a stereo macroscope at fixed magnification and using consistent grazing incidence illumination. A drawing stencil containing a series of holes of different diameters was placed over the photographed origins and the hole size selected which best matched the chosen level of

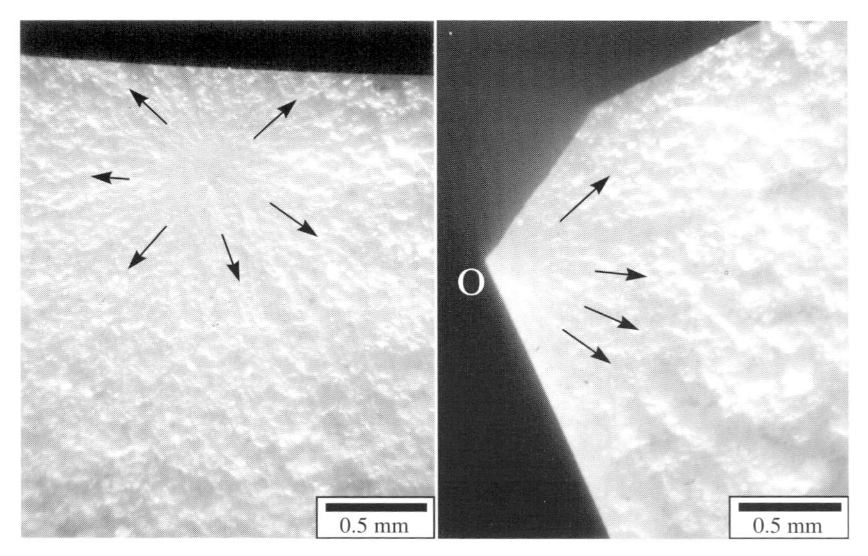

Figure 9. Examples of fracture origins in mechanically tested Y-TZP femoral heads, (a) internal origin from a pore, (b) surface origin ('O') at the end of the bore region adjacent to the cone edge chamfer. Arrows indicate directions of fracture ridges radiating from the origins.

contrast. To reduce possible observer bias, two observers made the measurements independently and the average result was taken. The effective mirror radius for each test-piece was then plotted against the flexural stress at the fracture origin (i.e. the nominal flexural stress corrected for distance of the origin from the outer test-piece surface) in order to determine the effective mirror constant for the material. Figure 10 shows that there is some scatter in

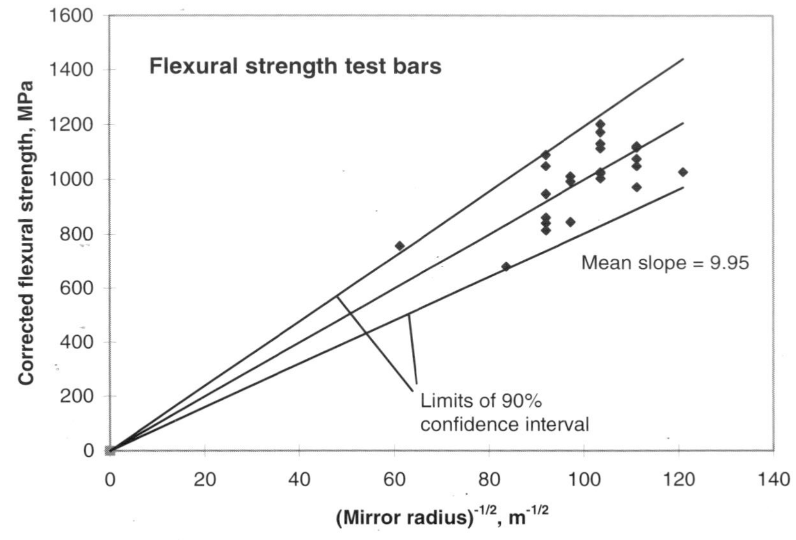

Figure 10. Plot of position-corrected flexural strength of Y-TZP test bars against 1/(mirror radius)$^{1/2}$. The central line, the average mirror constant, has a slope of 9.95 MPa m$^{1/2}$ and the other lines represent the limits of the 90% confidence interval.

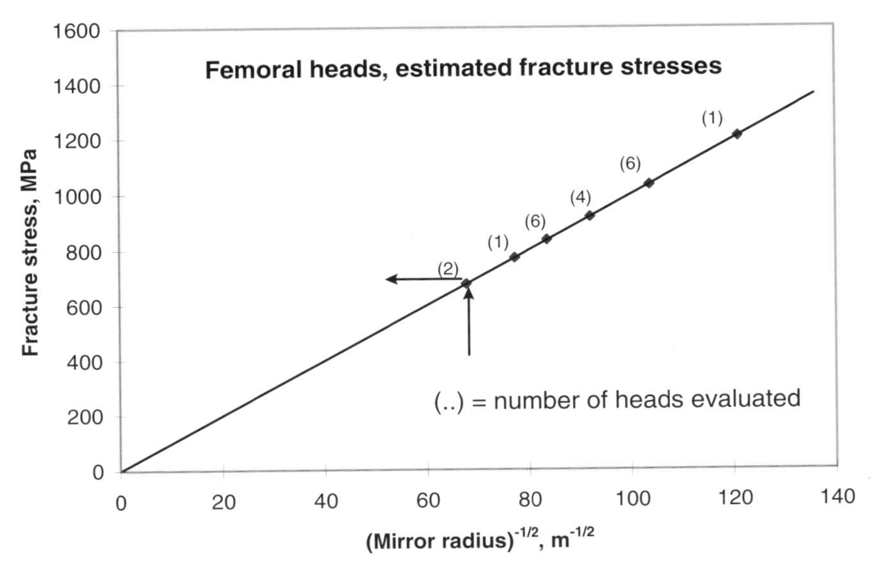

Figure 11. Fracture mirror plot for femoral heads, plotting fracture mirror radii onto the line representing the mean fracture mirror constant determined in Fig. 10, and permitting the fracture stress at the origins to be estimated.

data, which is not surprising in the circumstances, but the method permits a reasonable estimate of a mean value and the limits of the 90% confidence interval to be obtained.

The measurement procedure was repeated for the fractures in femoral heads. These data were then plotted on the line for the average mirror constant in order to determine the stress level at fracture. It can be seen in Fig. 11 that the fracture stresses thus determined are all in excess of 600 MPa, which is a respectable minimum value for Y-TZP in complex component form, compared with 1000 MPa average for small test bars. It could therefore be deduced that the ceramic material itself and its method of finishing were not the cause of reduced performance, so abnormally high localised stresses must have originated from the geometry of contact with the hard cobalt chrome alloy stem. After modification of the spigot geometry, the particular head design passed the minimum test performance requirements. Thus fractography showed this to be a system failure rather than a material or a design failure.

CONCLUSIONS

A review has been given of the application of fractography to advanced technical ceramics, illustrating the possibilities for identifying the underlying causes for the performance achieved by individual test-pieces or components as either intrinsic limitations in the material microstructure or extrinsic limitations caused by surface preparation, handling or use. The efforts being made to promote the use of fractography as an adjunct to material development, quality control, trouble shooting and in-service failure analysis have been outlined. An example from the orthopaedic field of application has been given to illustrate how fractography can assist in determining design or performance limitations to ensure safe use of ceramic components.

ACKNOWLEDGEMENTS

Part of this work has been supported by the Department of Trade and Industry through the Materials Metrology Programme

REFERENCES

1. E. Bourry: *A Treatise on Ceramic Industries*, 2nd edn, Scott, Greenwood & Son, London, 1911, 343.
2. R. W. Rice: in *Fractography of Ceramic and Metal Failures*, J. J. Mecholsky, and S. R. Powell eds, STP827 ASTM, Philadelphia, PA, 1984, 5–103.
3. J. D. Varner and V. D. Frechette (eds.): *Fractography of Ceramics and Glasses II: Ceramic Transactions 17*, 1988, American Ceramic Society, Westerville, OH.
4. R. C. Bradt and R. E. Tressler (eds.): *Fractography of Glass*, Plenum Press, New York, 1994.
5. J. D. Varner, V. D. Frechette and G. D. Quinn (eds.): *Fractography of Ceramics and Glasses III: Ceramic Transactions 64*, 1996, American Ceramic Society, Westerville, OH.
6. E. H. Yoffe: *Philos. Mag.* 1951, **42**, 739–750.
7. J. W. Johnson and D. G. Holloway: *Philos. Mag.* 1968, **17**, 899–910.
8. E. K. Beauchamp: in *Fractography of Ceramics and Glasses III: Ceramic Transactions 64*, 1996, American Ceramic Society, Westerville, OH, USA 409–445.
9. 'Standard practice for fractography and characterisation of fracture origins in advanced ceramics'. ASTM Standard C1322-96a, ASTM, W. Conshohocken, PA.
10. J. J. Swab and G. D. Quinn: 'Fractography of advanced structural ceramics: results from the VAMAS round robin exercise', VAMAS Technical Report No. 19, NIST, Gaithersburg, MD20899, USA, 1995.
11. 'Provisional test methods for fracture toughness of advanced ceramics at ambient temperature' ASTM PS 70–97 (now numbered as ASTM Standard: C1421-99), ASTM, W. Conshohocken, PA.
12. R. Morrell: *Fractography of Brittle Materials*, NPL Measurement Good Practice Guide No. 15, National Physical Laboratory, Teddington, 1999.
13. 'Implants for surgery – Partial and total hip joint prostheses – Part 5: Determination of resistance to static load of the head and neck region of stemmed femoral components', ISO Standard 7206-5:1992.
14. 'Implants for surgery – Partial and total hip joint prostheses – Part 6: Determination of endurance properties of head and neck region of stemmed femoral components', ISO Standard 7206-6:1992.

Polishing and Wear Behaviour of Sintered Alumina-Silicon Carbide Nanocomposites

A. M. COCK, H-Z. WU and S. G. ROBERTS
Department of Materials, Oxford University, Parks Road, Oxford, OX1 3PH, UK.

ABSTRACT

Alumina-silicon carbide nanocomposites have been observed to have polishing behaviour superior to that of pure alumina, although detailed studies of this effect are reported sparsely. This paper presents a systematic study of the polishing behaviour, under different conditions, of pressureless sintered nanocomposites made from powders which were prepared under near-industrial conditions. The polished surfaces were characterised by surface roughness measurements combined with reflected light microscopy. The results are compared with those from nanocomposites made from powders prepared under laboratory conditions. For both sets of samples, the polishing behaviour is improved compared to that of alumina. Microscopy clearly shows that the improved behaviour is due primarily to a change from intergranular fracture in alumina to predominantly transgranular fracture in the nanocomposites. This change is connected with the different subsurface deformation mechanisms as observed by transmission electron microscopy of polished nanocomposites.

1. INTRODUCTION

Much research has been done into understanding the so-called nanocomposite effect after Niihara's original paper[1] on Al_2O_3/SiC nanocomposites. One of the nanocomposite effects that has been observed is their improved polishing behaviour, compared to pure Al_2O_3[2,3]. In a recent paper, Kara and Roberts[4] performed a series of controlled polishing experiments at various grit sizes (150 μm flatbed grinding, followed by polishing with 25 μm, 8 μm and 3 μm diamond slurries) on samples produced by pressureless sintering under laboratory conditions. The surface finish at each stage of polishing was quantified by surface profilometry and by image analysis of reflected light micrographs to determine the relative amounts of smooth, reflective surfaces (transgranular fracture) and rough, non-reflective surfaces (grain pull-out and intergranular fracture). Their major conclusions were that (a) the polishing behaviour of Al_2O_3 is highly dependent on grain size, whereas there is little dependence on grain size in nanocomposites; and (b) nanocomposites with a comparable grain size to Al_2O_3 showed a smoother surface finish at all stages of polishing and could readily be polished to near 100% surface reflectivity whatever their grain size.

This paper presents another series of controlled polishing experiments, under the same conditions used by Kara and Roberts, on nanocomposites, pressureless sintered from powders produced by a 'near industry standard' route, i.e. that of spray drying and uniaxial pressing only, and compares the results to Kara and Roberts' experiments. In addition, by studying the subsurface deformation by using transmission electron microscopy (TEM) of cross-sections of polished alumina and nanocomposites, some understanding of the different polishing behaviours observed is obtained.

2. SAMPLE PREPARATION

Table 1 summarises the powders used in the production of the Al_2O_3 and Al_2O_3/SiC nanocomposites used here and previously by Kara and Roberts. Both the Al_2O_3 powders are available in commercial volumes; however, the one major difference between them is that the AES11C has been wet ball milled, whereas the DBM 172 powder was dry ball milled by the vendors. The dry ball milled material is more broken up and contains less agglomerates, which aids subsequent powder processing and results in fewer flaws after sintering. The two silicon carbide types were received from Lonza, Germany. The UF45 has a slightly smaller mean particle size and narrower distribution (D_{10}=0.18, D_{50}=0.35, D_{90}=1.14) than the UF25 (D_{10}=0.21, D_{50}=0.43, D_{90}=1.14), but it is not available in commercial quantities.

Full details of powder preparation routes are available elsewhere[5]. The major differences between the two processing routes are shown in Table 1. Obviously the 'industrial' route used here is designed to produce larger quantities, and for cost reasons, precludes the use the 'laboratory methods' of vacuum and freeze drying. The binder addition aids the formation of the green bodies hence removing the need for cold isostatic pressing, removing another cost factor, and aiding the reduction in sintering temperature, which is also of commercial significance.

The results of the sintering experiments are shown in Table 2. The materials are all close

Table 1. Differences in powder preparation for industrial scale (this paper) and laboratory scale (Kara and Roberts) nanocomposite production.

	This study	Kara and Roberts
Al2O3	Pechiney DBM 172	Sumitomo AES11C
SiC	Starck UF25	Starck UF45
Lot size	~ 50 kg	~ 1 kg
Milling	Ball	Attrition
Binder	Yes	No
Drying	Spray	Vacuum and Freeze
Green bodies	Uniaxial pressing only	Uniaxial and cold isostatic pressing
Sintering	Flowing N2 on SiC bed.	Flowing N2 on SiC bed.
Sintering temperature	Al2O3: 1550-1650°C Al2O3/SiC: 1550-1650°C	Al_2O_3: 1550-1600°C Al_2O_3/SiC: 1600-1700°C

Table 2: Properties of the Al_2O_3 and Al_2O_3/SiC nanocomposites.

	Sintering temperature °C	Density g cm^{-3}	Density %theoretical	Mean grain size (μm)
Al2O3	1550	3.937	99.30	5.18
	1650	3.911	98.64	6.71
1.25 vol% SiC	1550	3.912	98.67	4.14
	1650	3.923	98.95	4.34
4.9 vol% SiC	1550	3.746	95.19	3.13
	1650	3.860	98.09	2.98

to full density, except in the case of the 4.9 vol% SiC nanocomposites sintered at the lower temperatures. The effect of increased silicon carbide content on the grain size is parallel to that found by Kara and Roberts in that the presence of SiC restricts grain growth. Kara and Roberts found that 1% SiC and a sintering temperature of 1700°C had little effect on the grain size, although, in this study 1.25% SiC appears to have a limited effect at sintering temperatures of 1550°C and 1650°C. Kara and Roberts explored the grain size effect on polishing, also observed by Marshall et al.[6], as their Al_2O_3 grain size ranges from 2.0-5.3 μm and the Al_2O_3/SiC grain size ranges from 1.4 – 6.5 μm. The Al_2O_3 samples prepared here have a larger grain size, 5.2-6.7 μm, than that of Kara and Roberts, whilst the Al_2O_3/SiC samples have a grain size ranging from 3.0 –4.3 μm, enabling further investigation of this grain size effect.

3. POLISHING EXPERIMENTS

The polishing procedure is outlined below and was similar to that used by Kara and Roberts. Five samples were mounted on steel disks at a fixed radius as shown in Fig. 1 and ground using a flat bed grinder (Model 1400L, Jones and Shipman, Leicester, U.K.) under the following conditions: 250 mm diameter resin-bonded 150-grit diamond wheel; grinding wheel speed of 1240 rpm; table velocity of 0.8 ms^{-1}; and depth of cut 0.125 mm/pass.

After grinding, polishing was performed on a Logitech PM2 polishing machine with a surface pressure of 0.01 MPa. Three different grades of diamond slurry, 25, 8 and 3 μm grit size, were used successively. For the 25 and 8 μm grit sizes, the base plate was cast iron and the total polishing time was 45 minutes at each grit size. For the 3 μm grit size, a copper base plate was used and the total polishing time was 90 minutes. The diamond slurry was automatically fed using a Kemet dispenser (2T Series 3 CE EMC Dispenser, Kemet International Ltd., Maidstone, U.K.) under the same conditions for each stage.

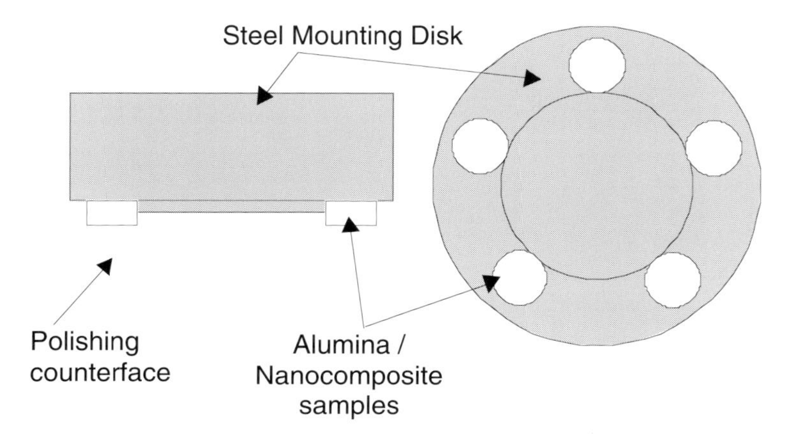

Fig. 1 Sample mounting for controlled polishing tests. The diameter of the steel disk is 125 mm. The diameter of each sample is 20 mm, and the sample centres are mounted on a diameter of 100 mm.

After each polishing stage, a reflected light micrograph of the sample surface was taken and surface profilometry measurements were made using a Surtronic 3+ Stylus Profilometer (Rank Taylor Hobson Ltd., Leicester, U.K.). At each polishing stage, and on each sample, twenty measurements, 4 mm in length, were made – ten parallel to and ten perpendicular to the grinding direction except in the case of the samples with a 3 μm polish where five measurements were made in each direction. For the ground samples, a small difference between the parallel and perpendicular R_a values was measured, but this disappeared after the first polishing stage.

4. RESULTS AND DISCUSSION

4.1. *Progressive polishing*

Figures 2 and 3 show reflected light micrographs of the samples, pressureless sintered at 1650°C and 1550°C respectively, at the sequential polishing stages. The micrographs for the materials made at the two different sintering temperatures are very similar at each polishing stage showing that the slight differences in density have little effect on the polishing behaviour.

The alumina samples show very poor reflective surfaces, indicating much intergranular fracture and grain pullout at all stages, until the 3 μm polish where there is a significant amount of polished surface but also with some grain pullout. The 1.25 vol% SiC samples show a similar surface to the alumina in the ground state but at both the 25 and 8 μm stages there are grooves present at various angles indicating that there is plastic deformation occurring at these stages. A highly smooth surface is observed after the 3 μm polish. For the 4.9

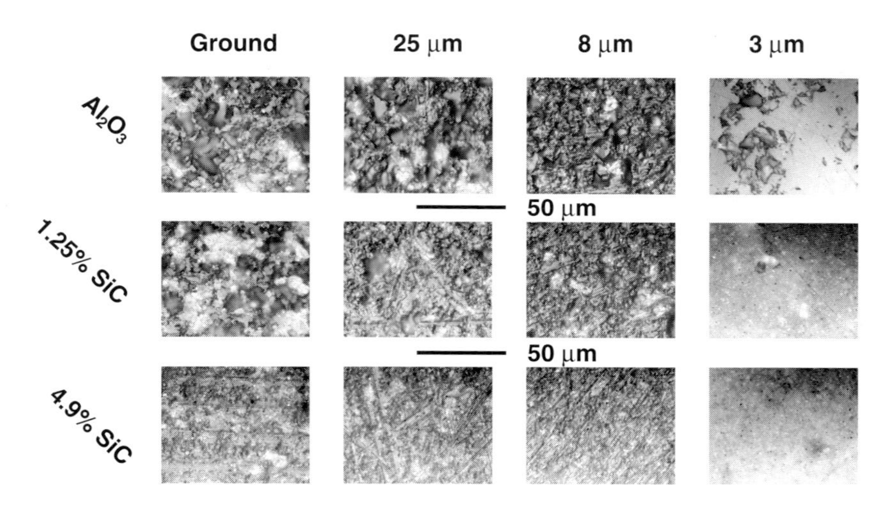

Fig. 2 Reflected light micrographs of the 1650°C pressureless sintered samples at the various stages of polishing.

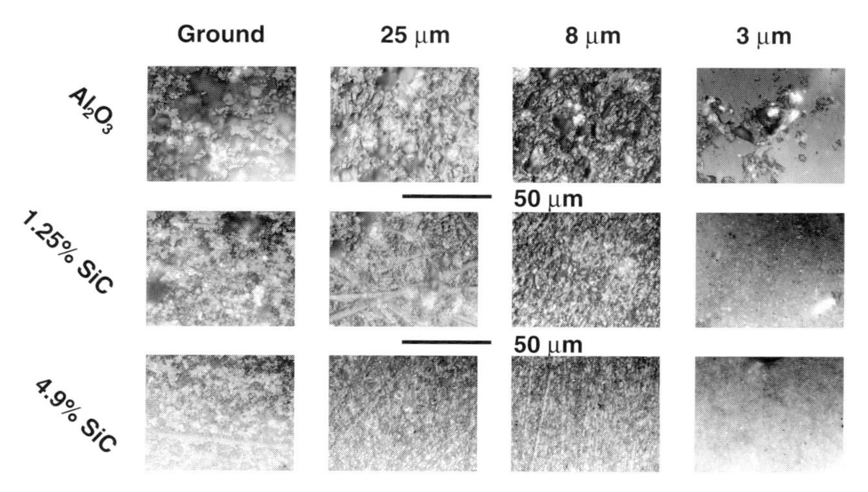

Fig. 3 Reflected light micrographs of the 1550°C pressureless sintered samples at the various stages of polishing.

vol% SiC samples grooves are also observed in the ground sample as well as in the 25 and 8 μm polishing stages and again a smooth surface is obtained at the 3 μm stage.

The surface profilometry measurements are shown in Figure 4. They confirm the trend shown by the micrographs in that the alumina polishes worst (with the most grain pullout and intergranular fracture). The 1.25 vol% SiC nanocomposites have lower R_a values than

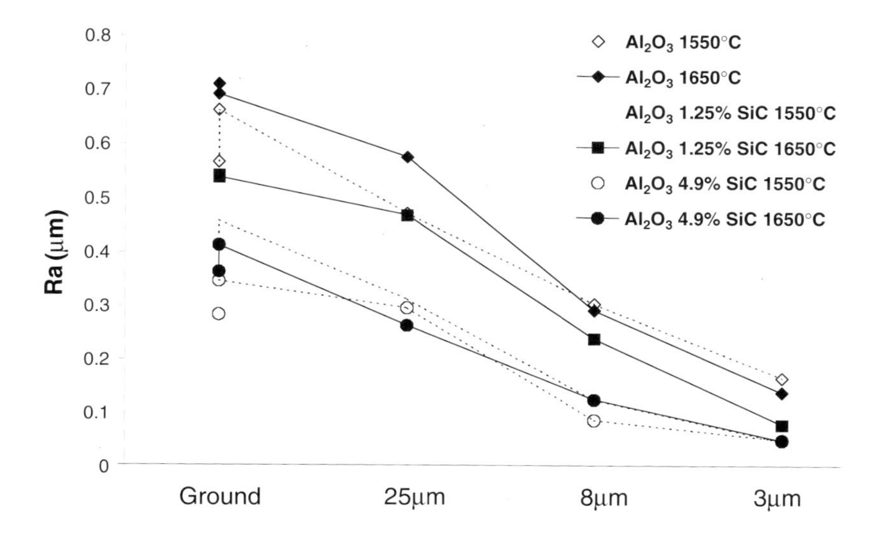

Fig. 4 Surface profilometry results (R_a values) from the polishing experiments. The ground surface measurements show R_a values parallel and perpendicular to the grinding direction. The lower value in each case is for the 'parallel' measurement except in the case of the 1650°C Al2O3.

the alumina at each stage, probably because of the increased amount of transgranular fracture, and end with a smooth surface finish at the final polishing stage. Finally, the 4.9 vol% SiC nanocomposite produces the smoothest surfaces at each polishing stage. This is also in agreement with Kara and Roberts' findings.

4.2. *Grain size*

Figure 5 shows the surface roughness measurements at the final polishing stage versus grain size for the materials prepared here and for those of Kara and Roberts. The final surface finish of Kara and Robert's alumina specimens is clearly dependent on grain size, whereas their nanocomposites show little dependence of surface finish on grain size. The alumina prepared in this study does not behave in the same way as Kara and Robert's alumina, which may be due to the different alumina powder used. The nanocomposites prepared here do not polish to quite the same surface finish as Kara and Robert's samples, and there is also no apparent grain size dependence. The graph highlights the low vol% SiC nanocomposites, as they have a relatively large range in grain size yet still polish to an equally good surface finish, indicating that the SiC in the nanocomposites is more important than grain size in determining surface finish.

4.3. *Deformation mechanisms*

Figure 6 shows two transmission electron micrographs of the cross-sections through the ground surfaces of alumina and a 5 vol% SiC nanocomposite. These samples were manufactured by hot pressing to produce fully dense material to facilitate specimen preparation for the TEM. These materials have a similar microstructure to the materials presented here. In response to grinding, the alumina deforms by twinning with some slip present,

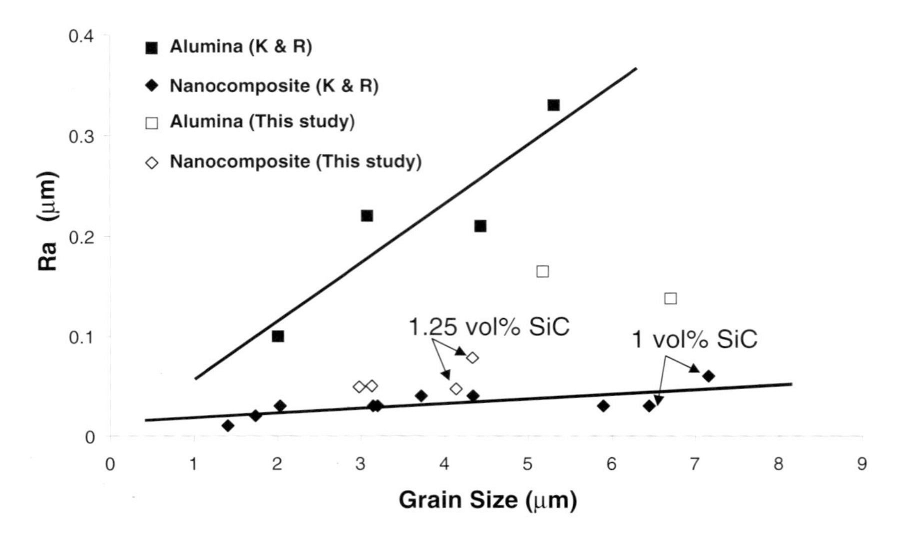

Fig. 5 Surface roughness dependence on grain size after the final (3 μm) polishing stage.

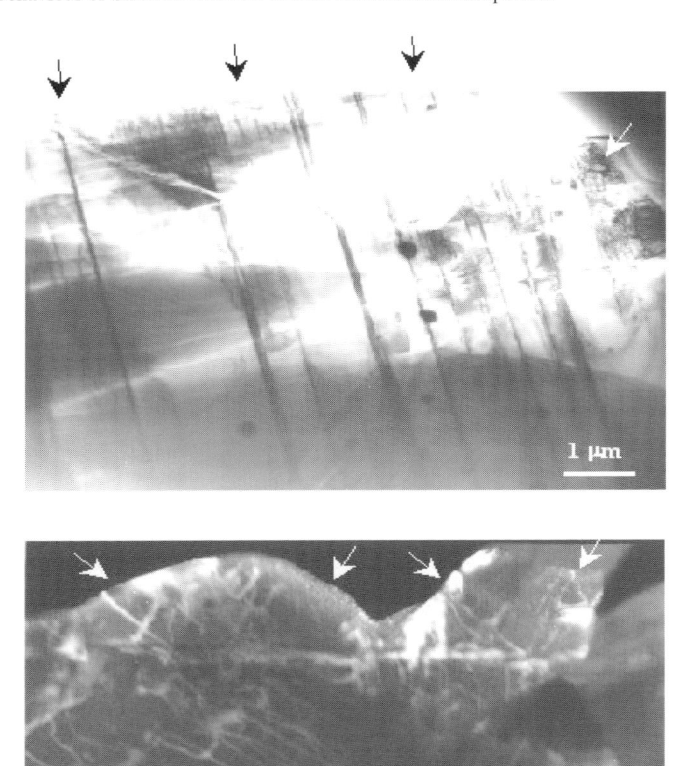

Fig. 6 Transmission electron micrographs of cross-sections through (a) the ground surface of hot pressed alumina and (b) the ground surface of a hot pressed 5 vol% SiC nanocomposite. The arrows depict the top surface. The dominance of twins within the alumina and dislocations within the nanocomposite is clearly evident.

whereas the nanocomposite only shows slip. TEM of polished samples has shown that they deform only by slip. The major difference between the two surface treatments is the depth below the surface to which dislocations are observed. For a ground alumina surface, the dislocation substructures are limited to the surface layer of grains and to a depth of less than 1.0 μm but in the nanocomposite this depth is 3.0-10 μm, i.e. two or three grain diameters. For a 1 μm polish the dislocation depth is less than 0.3 μm in alumina and in the nanocomposite it is in the range of 0.5-1.5 μm. More details are presented by Wu et al[7].

These different dislocation structures may be due to either (a) the thermal mismatch between the SiC and Al_2O_3 generating local dislocation sources on cooling after sintering which may assist plastic deformation when external stresses are applied, or (b) the alumina having weaker grain boundaries than the nanocomposites, and so never developing the more complex dislocation structure as grain pull out occurs first.

5. CONCLUDING REMARKS

The nanocomposites prepared by a near-industrial route have shown very similar polishing behaviour to nanocomposites prepared in the laboratory. There was no grain size effect on their polishing behaviour and all the nanocomposites could be polished to a good surface finish. The deformation mechanism observed in the polishing of the alumina and nanocomposites shows a change from predominantly intergranular fracture to a mixture of transgranular and intergranular fracture. This change is not fully understood, but, as the TEM observations of Wu et al. have shown, this change may be connected with the underlying deformation mechanism of the material. There is some suggestion that the grain boundaries in the nanocomposites are stronger than in alumina[8] due to the formation of SiO_2 or an SiO_2/Al_2O_3 compound such as mullite[9]. The critical information that has yet to be obtained relates to the nucleation and initial growth of cracks in these materials. This work is ongoing.

ACKNOWLEDGEMENTS

We are grateful to the European commission for supporting this work under a Brite Euram III Project: BRE3 CT96 0212 and particularly to Morgan Matroc, one of our Brite Euram partners, for the preparation of alumina and alumina-silicon carbide powders.

REFERENCES

1. K. Niihara and A. Nakahira: *J. Ceram. Soc. Jpn.*, 1991, **99** [10], 974-982.
2. A. J. Winn and R. I. Todd: *Br. Ceram. Trans.*, 1999, **98** [5], 219-224.
3. A. J. Winn and R. I. Todd: *J. Eur. Ceram. Soc*, In press.
4. H. Kara and S. G. Roberts: *J. Am. Ceram. Soc*, In press.
5. R. W. Davidge, R. J. Brook, F. Cambier, M. Poorteman, A. Leriche, D. O'Sullivan, S. Hampshire, and T. Kennedy: *Br. Ceram. Trans.*, 1997, **96** [3], 121-127.
6. D. B. Marshall, B. R. Lawn and R. F. Cook: *J. Am. Ceram. Soc*, 1987, **70** [6], C139-C140.
7. H.-Z. Wu, S. G. Roberts and B. Derby: *J. Am. Ceram. Soc*, submitted.
8. I. Levin, W. D. Kaplan, D. G. Brandon and A. A. Layyous: *J. Am. Ceram. Soc*, 1997, **17**, 921-928.
9. H.-Z. Wu, S. G. Roberts, A. J. Winn and B. Derby: *Proc. MRS 1999 Fall meeting*, In press.

Ink Jet Printing of Alumina Suspensions in Liquid Wax

N. REIS[†‡], K. A. M. SEERDEN[†], P. S. GRANT[†],
B. DERBY[‡], AND J. R. G. EVANS*
†Department of Materials, University of Oxford, Parks Rd., OX1 3PH, U.K.
‡Manchester Materials Science Centre, UMIST, Grosvenor Street, M1 7HS, UK.
**Department of Materials, Queen Mary and Westfield College, University of London, E1 4NS, UK.*

ABSTRACT

This paper reports on our recent achievements, using a hot-melt ink system, towards the fabrication of ceramic bodies by controlled droplet deposition. Key factors for jetting were identified and the rheological behaviour of fairly concentrated particulate suspensions was optimised to match the window of print-head operation. As a demonstrator, several unfired objects containing 30 vol.% colloidal Al_2O_3 dispersed in paraffin were fabricated, using a three-dimensional ink jet plotter primarily developed for rapid prototyping. This platform is currently being improved to allow the fabrication of shapes with higher particulate contents. Preliminary studies using a 75 μm orifice diameter ink jet print-head have produced encouraging results.

1. INTRODUCTION

Today's ink jet technology (IJT) extends its application much beyond that of Desktop Publishing. Although the latter has been the driving force for its development, ink jet processes are now seen as versatile, fully automated, and relatively inexpensive ways for depositing droplets in a reproducible and precise fashion. Recent applications include the fabrication of DNA chip micro-arrays,[1,2] solder bumping for flip-chips[3] and printed circuit boards,[4] optical micro-lenses,[5,6] and functionally graded drug delivery devices.[7]

Perhaps most attractive from a manufacturer's perspective, is the fusion of IJT with computer solid modelling to realise desktop micro-fabrication. Three-dimensional printers introduced for prototyping and fabrication of casting masters and moulds, are stimulating achievements in this area.[8-10] The layer-wise strategies employed by the above allow rapid realisation of optimally designed 3D structures which would be either difficult or impossible to process by conventional machining and casting techniques.

Upgrading IJT for flexible manufacturing presents the materials community with the familiar challenges for formulating and compounding raw materials to meet the processing requirements. Direct ink jet deposition of ceramic suspensions is no exception and the compromise that needs to be established between ink rheological properties for jetting, and ceramic firing ability is paradigmatic. Earlier work on commercially available 2D ink jet printers used ceramic inks with low solids contents (typically less than 10 vol%) and evaporative carriers.[11-13]

Building thicker sections with better feature definition within reasonable time scales demands inks with higher particulate concentrations. In this regard, phase change piezoelectric ink jet printing appears to be an interesting alternative. It is less demanding in terms of ink properties (such as conductivity required for continuous ink jet or high vapour pressure for thermal ink jet), and allows compensation for viscous losses due to powder loading using higher driving pressures and/or temperature adjustments. Moreover, drop formation is easily controlled by manipulating the actuator excitation pulse, and it produces smaller splats on impact due to droplet freezing.[14-16]

According to our experience with tubular piezoelectric hot-melt print heads, the viscosity upper limit for jettable fluids lies between 40 and 50 mPa s. Concentrated suspensions having solids contents between 40 and 50% and such low viscosity are difficult to prepare but possible to obtain, provided suitable carriers and dispersing systems are used.[17,18] Moreover, strategies employing suspending phases that can, under controlled conditions, be removed or modified after jetting may also be envisaged.

2. EXPERIMENTAL PROCEDURE

2.1. *Suspension formulation*

A commercial grade α-Al_2O_3 powder (RA45E – Alcan Chemicals) was used in this study. The powder has a mean particle size of 0.4 μm (d_{90} = 1.7 and d_{10} = 0.2 μm), and a specific surface area of 8 $m^2.g^{-1}$. For the suspending medium a paraffin wax (MobilWax135™ – Mobil Special Products) was used, since it is a common ingredient in various powder forming methods. This paraffin has a melting point of 57°C, and viscosity and density at 100°C of 2.8 mPa s and 780 kg m^{-3}, respectively.

To stabilise the suspensions a range of proprietary dispersants and their combinations with either stearic acid (1-octadecanoic acid, BDH Chemicals) or sterylamine (1-octadecylamine, Lancaster Synthesis, U.K.) were tried. The commercial surfactants Hypermer FP1 and LP1 (both from ICI Surfactants) gave the best results for this system. According to the manufacturer, these particular two dispersants are polydisperse organic compounds containing carbonyl terminated groups,[19] differing solely in molecular weight.

2.2. *Suspension preparation and characterisation*

Suspensions were prepared by melting the wax and subsequently mixing the different dispersing agents under gentle stirring. These mixtures were poured into HDPE bottles containing alumina beads before adding the powder. The resulting slurries were ball milled inside an oven at 120°C for 10 hours. These suspensions were filtered using a stainless steel wire mesh (500 mesh count – 25 μm wire diameter) and then allowed to solidify to room temperature.

Rheological characterisation was performed on a concentric cylinder rheometer (RS III - Brookfield), equipped with a re-circulating system for high temperature evaluation. Steady shear measurements were performed at shear rates between 3 and 300 s^{-1} by incrementing the shear rate and measuring the shear stress after an equilibration period (typically between

30 and 60 s). The shear rate range was limited by the shear stress developed by the suspension, that range decreasing with increasing particulate content (or increasing viscosity).

2.3. *Jetting evaluation and printing trials*

In order to determine the optimum ink jet printing parameters for each individual suspension a jet monitoring station was set up. The apparatus uses the same driving electronics of a commercial 3D ink jet plotting system (ModelMaker 6PRO, Sanders Prototype Inc.), modified to control independently each parameter and trigger an illuminating light emitting diode at fixed or variable time delays from the actuator firing pulse. Details on the experimental set-up can be found elsewhere.[20] Stroboscopic imaging and recording of the droplets near the ink jet head were performed using a conventional CCD camera and a PC frame grabbing card. All printing heads used have a nozzle orifice diameter of 75 μm. Quantities such as average droplet velocity and volume can easily be extracted from image analysis and simple weighing methods.

Printing trials with optimised ceramic inks were carried in a ModelMaker 6PRO, by replacing reservoirs, heated lines and jet heads to avoid contamination. For the demonstrator objects, pulse amplitude (actuator excitation voltage), temperature and height of the reservoir relative to the printing head were varied. The milling system used to flatten the top of each layer and establish the layer thickness was disarmed to avoid smearing of the deposit and consequent loss of detail.

3. RESULTS AND DISCUSSION

Colloidal stability and steady shear properties

From all dispersants tested it was found that Hypermer LP1 combined with sterylamine (SA) provided the best results in terms of dispersing effectiveness and hence lowest viscosity. Earlier work has shown that Hypermer FP1 produce slightly higher viscosities,[21] which is assumed to be related to the lower molecular weight (the only reported difference between FP1 and LP1 according to the manufacturer) but the mechanism for this difference is not certain.

Qualitative sedimentation experiments have shown that combining short chain amine terminated dispersant with polymeric carbonyl side chain dispersants increase the stability of the alumina suspended paraffins. FTIR experiments have shown that both groups attach at the particle surface suggesting amphoteric behaviour from the alumina.[22] Similar observations were also reported for SiC suspended in wax using the same fatty amine and an alkylsuccimide.[23]

A third observation was that utilising two dispersants with different molecular weights generally resulted in better dispersions, than using either of them separately. This effect has also been observed by other researchers with systems based on paraffins and similar surfactants and is attributed to the more extended configuration of the adsorbed molecules and hence denser and thicker stabilising layers.[23,24]

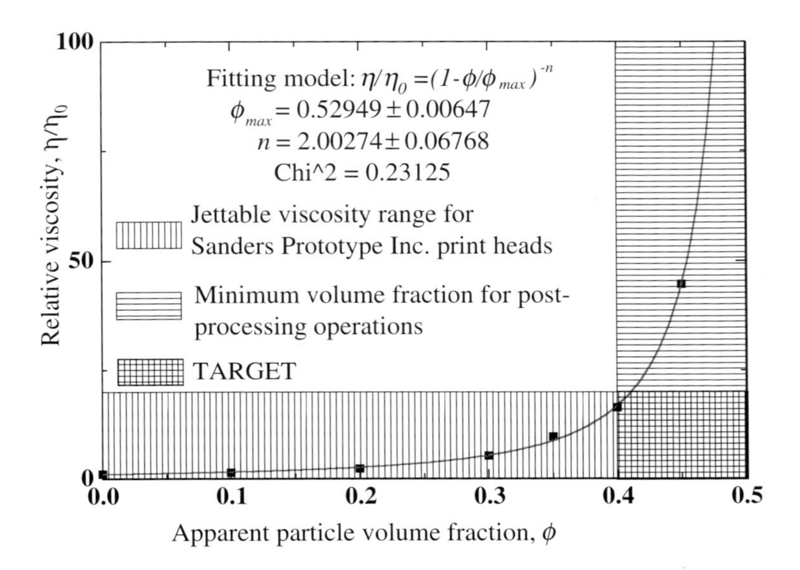

Fig. 1 Relative viscosity against volume fraction. Steady shear measurements performed at a shear rate of 100 s^{-1} and a temperature of 120°C. Volume fraction does not take the steric stabilisation layer into account and therefore is denoted as apparent. (Chi^2 is the least squares error coefficient)

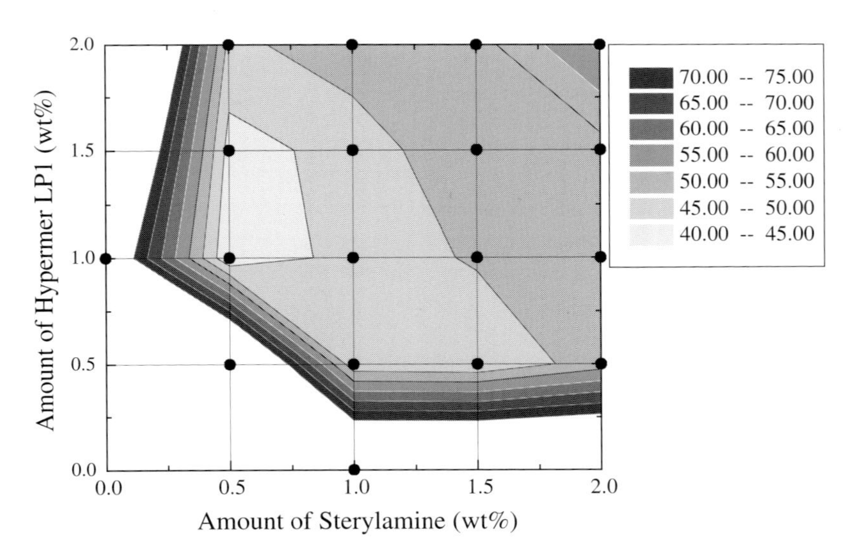

Fig. 2 Steady shear viscosity contour map for a 40 vol% suspension as a function of dispersant concentration (based on the weight of the dry powder). Measurements performed at shear rates of 80 s^{-1} and temperatures of 120°C, for the points indicated (•). The units on the legend are mPa s and the white region accounts for viscosities greater than 75 mPa s.

From the above a systematic investigation of dispersing systems containing SA and LP1, for high particulate contents suspensions exhibiting low viscosity was carried out. Figure 1 shows the relationship between viscosity and particle volume fraction, fitted to a modified Krieger-Dougherty model.[25] The fitting parameter ϕ_{max}, which accounts for the approach to infinite viscosity, is somewhat lower than that predicted by theory. This result has been explained by the increase in particle volume due to the presence of the steric stabilising layer, which should be used for calculating the effective volume fraction.[17] Nevertheless, the value is somewhat lower than expected, meaning that the ϕ_{max} asymptote can possibly be shifted toward higher particulate contents, and thus extend low viscosities to higher volume fractions.

A systematic variation of the amounts of SA and LP1 was undertaken in order to determine the ratio between the two yielding lowest viscosity. The results are shown in Fig. 2 as a steady shear viscosity contour map for a 40 vol% suspension. It can be seen that weight ratios of 1:2 to 1:3 (larger to smaller surfactant molecule) provide lower viscosities. It is self evident, that the excess of surfactant molecules, particularly LP1, is detrimental to the viscosity due to the polymeric nature of the dispersant. This result was confirmed by viscosity measurements carried on unfilled systems, showing that equivalent increments of 1wt% in the amount of LP1, increased the viscosity by 15 to 20%.

Jetting and printing results

Typical recording images obtained with the monitoring station are shown in Fig 3. Due to the resonant nature of piezoelectric actuated ink jets, drop volume and velocity can be controlled by the excitation pulse shape and frequency. It was earlier reported that increasing

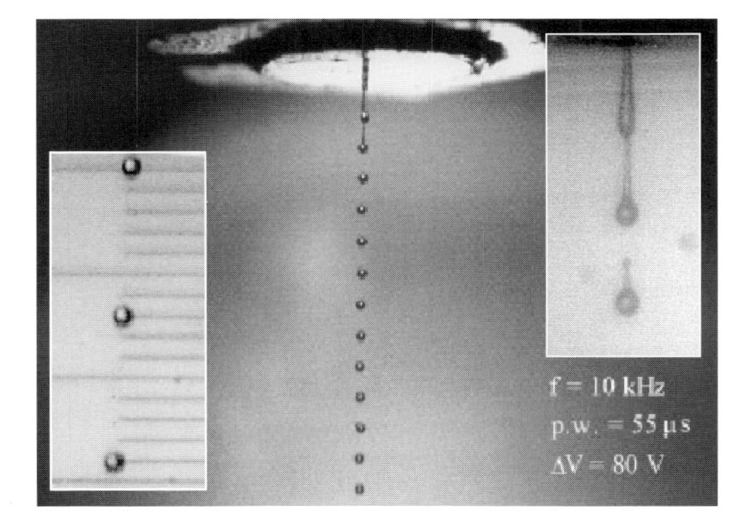

Fig. 3 *In situ* stroboscopic imaging of the forming droplets. Material is a blend of fatty esters and fatty acids with a melting temperature range of 60–70°C, density of 930 kg m^{-3}, and viscosity and surface tension at 120°C of 8 mPa s and 26.5 mJ.m^{-2}, respectively. On the high magnification pictures the minor divisions of the scale are 100 μm. Drop radius and velocity are 40 μm and 7 m s^{-1}.

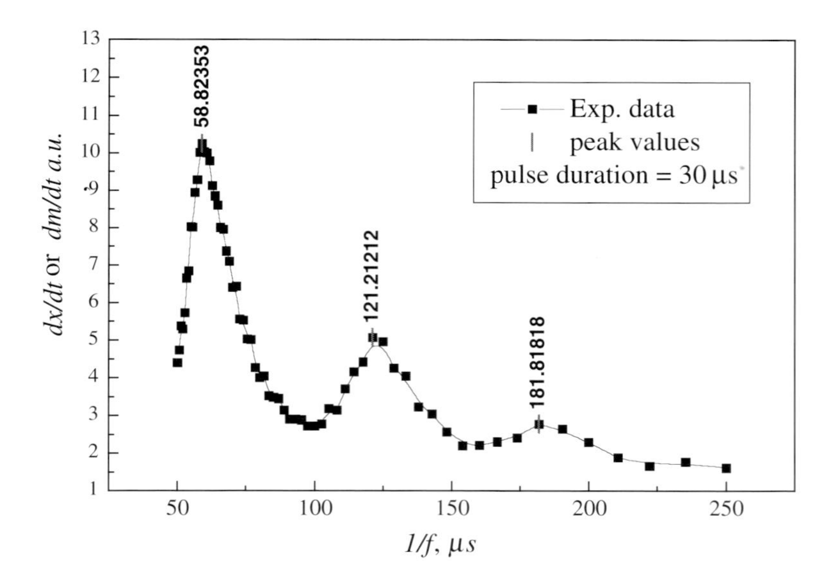

Fig. 4 Mass rate (or drop velocity) as a function of firing pulse repeat time at high frequencies (arbitrary units). Peaks and valleys indicate sub-harmonic resonance and anti-resonance characteristics of the fluid chamber depending on the interference between reflecting acoustic waves and the transducer firing frequency. Fluid used is the same as shown in Fig 3.

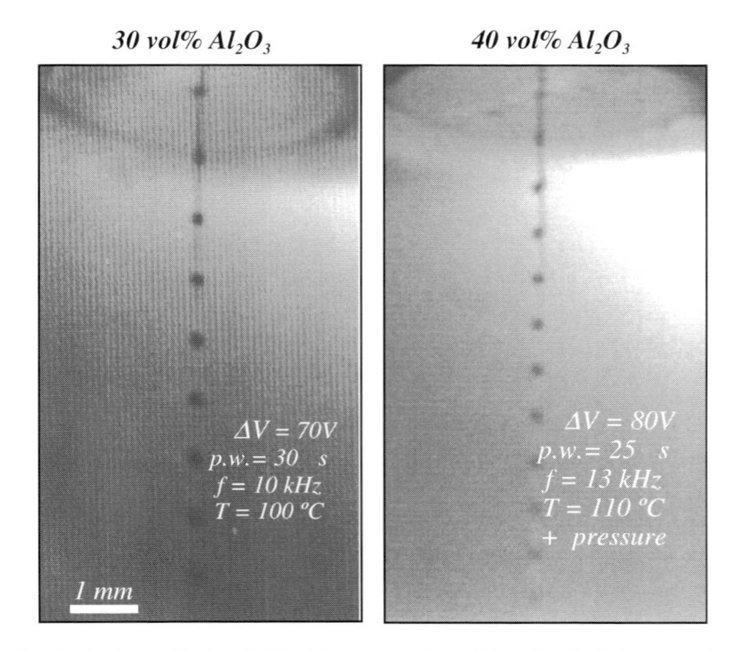

Fig. 5 Jetting behaviour of 30 and 40 vol% suspensions. Note that high frequencies are required to jet these fluids and a pressure is needed to force the flow through the feeding lines to the jet head, for the 40 vol% suspension.

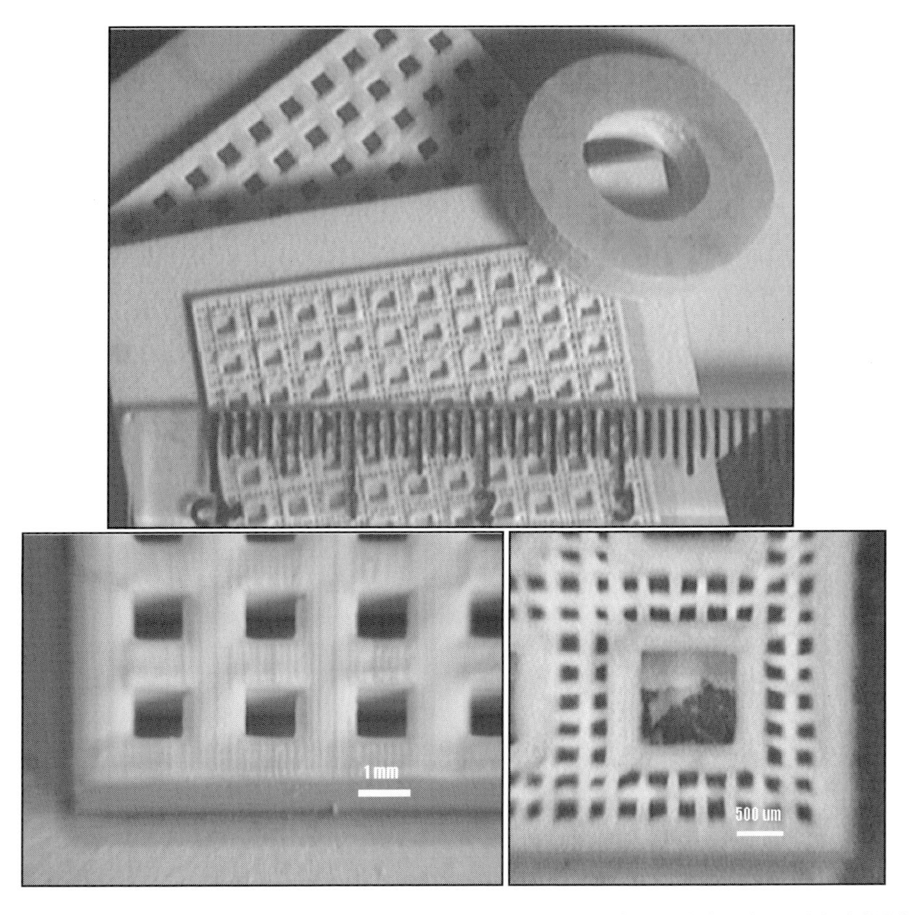

Fig. 6 Printed green ceramic objects with a particulate content of 30 vol.%, using a ModelMaker 6Pro, slightly modified to allow jetting of ceramic suspensions.

pulse amplitude results in higher droplet velocities, and there is an optimum pulse duration for maximum drop ejecting velocity. This is due to a resonance and hence depends on the chamber geometry and fluid acoustic properties.[20,26]

Although for a given fluid at constant temperature and low frequencies drop formation is mainly affected by pulse duration and amplitude, the situation changes when the jet heads are operated in the high frequency regime (typically above 6 kHz). In this case, the reflecting waves generated by one pulse are not extinct before the next pulse is applied, introducing sub-harmonic resonance and anti-resonance. This effect can be used to jet higher viscosity fluids at a given voltage and pulse width, by tuning the firing frequency for maximum constructive interference, as shown by the peaks in Fig 4. In fact, high frequencies were needed to jet concentrated suspensions – see Fig. 5.

Printed ceramic bodies with a solids content of 30 vol% are shown in Fig. 6. Driving parameters such as voltage and temperature were modified to avoid wetting of the nozzle face and consequent irregular jetting. For the 40 vol% suspension, the flow from the reservoir to the jet head had to be aided by a small amount of pressure, due to the low shear and hence highly viscous flow through the heated lines.

4. CONCLUSIONS

Phase change ceramic inks exhibiting low viscosities at high particulate concentrations can be achieved by choosing suitable steric stabilising dispersants. For alumina suspended in paraffin, combinations of surfactants with both acidic and basic functionalities, and different molecular weights yielded the lowest viscosity suspensions. Jetting of high viscosity fluids can be achieved by operating the jet head at very fast repetition rates and tuning that frequency for maximum constructive interference.

ACKNOWLEDGEMENTS

We would like to acknowledge the support of the EPSRC through grant GRL42537 and the support of the Portuguese Government of NR for his PhD funding through program PRAXIS XXI. We would also like to thank Sanders Design International for assistance with the design of our ink jet monitoring equipment.

REFERENCES

1. T. Goldmann and J. S. Gonzalez, *J. Biochem. Bioph. Meth.*, 2000, **42**, (3), 105-110.
2. B. Lemieux, A. Aharoni, and M. Schena, *Mol. Breeding*, 1998, 4, (4), 277-289.
3. A. F. J. Baggerman and D. Schwarzbach, *IEEE T. Compon. Pack B*, 1998, **21**, (4), 371-381.
4. D. J. Hayes, W. R. Cox, and M. E. Grove, *J. Electron. Manuf.*, 1998, 8, (3-4), 209-216.
5. R. Danzebrink and M. A. Aegerter, *Thin Solid Films*,1999, **351**, (1-2), 115-118.
6. S. Biehl, R. Danzebrink, P. Oliveira P, and M. A. Aegerter, *J. Sol-Gel Sci. Tech.*, 1998, 13, (1-3), 177-182.
7. B. M. Wu, S. W. Borland, R. A. Giordano, L. G. Cima , E. M. Sachs, and M. Cima, *J. Cont. Release*, 1996, **40**, (1-2), 77-87.
8. E. M. Sachs, J. S. Haggerty, M. J. Cima, and P. A. Williams, U.S. Patent N. 5,204,055, 20 April 1993.
9. R. C. Sanders JR., J. L. Forsyth, K. F. Philbrook, U.S. Patent N. 5,506,607, 9 April 1996.
10. R. N. Leyden and C. W. Hull, U.S. Patent N. 5,855,836, 5 January 1999.
11. M. Mott, J.H. Song, and J. R. G. Evans, *J. Am. Ceram. Soc.*, 1999, **82**, (7), 1653-1658.
12. S. J. Kim and D. E. McKean, *J. Mater. Sci. Lett.*, 1998, **17**, (2), 141-144.
13. G. Thornell, L. Klintberg, T. Laurell, J. Nilsson, and S. Johansson, *J. Micromech. Microeng.*, 1999, 9, (4), 434-437.
14. S. Schiaffino and A. A. Sonin, *Phys. Fluids*, 1997, **9**, (11), 3172-3187.
15. F. Q. Gao and A. A. Sonin *P. Roy. Soc. Lond A*, 1994, **444**, (1922), 533-554.
16. R. Bhola and S. Chandra, *J. Mater. Sci.*, 1999, **34**, (19), 4883-4894.
17. L. Bergström, *J. Am. Ceram. Soc.,* 1996, **79**, (12), 3033-40.
18. T. N. Tiegs and D. E. Wittmer, U.S. Patent N. 5,456,877, 10 October 1995. 1. J. D. Schofield, Eur. Patent N. 0240160, 17 April 1991.
19. N. Reis, K. A. M. Seerden, B. Derby, J. W. Halloran, and J. R. G. Evans, *Mater. Res. Soc. Symp. Proc.*, 1999, 542, 147-152.

20. K. A. M. Seerden, N. Reis, B. Derby, J. W. Halloran, and J. R. G. Evans, *Mater. Res. Soc. Symp. Proc.*, 1999, **542**, 141-152.
21. K. A. M. Seersden, unpublished work.
22. R. Lenk, *Ceram. Trans.*, 1995, 51, 303-306.
23. R. Lenk, A. G. Kriwoscepov, and J. G. Frolov, *Sprechsaal*, 1992, **125**, (12), 805-814. (in German)
24. L. Bergström, *J. Mater. Sci.*,1996, 31, (19), 5257-5270.
25. T. W. Shield, D. B. Bogy, and F. E. Talke, *IBM J. Res. Develop*, 1987, **41**, (1), 96-110.

Preparation of Zirconia Inks for Continuous Jet Printing

H. RASHID, B. Y. TAY and M. J. EDIRISINGHE
Department of Materials, Queen Mary and Westfield College, University of London, Mile End Road, London E1 4NS, UK

ABSTRACT

This paper describes the preparation of zirconia inks by triple roll milling and triple roll milling followed by ultrasonic disruption for use in a continuous ink-jet printer. The sedimentation behaviour, viscosity, surface tension and conductivity of the zirconia inks prepared have been measured. Ultrasonic disruption improved the dispersion of zirconia in the ink. However, some zirconia agglomerates were still present and these had to be removed by sedimentation prior to successful printing of a 2.5 vol.% zirconia ink.

1. INTRODUCTION

In the last two decades, forming processes for engineering ceramic materials has progressed beyond the traditional methods of powder compaction and slip casting. More recently, a class of new forming methods, known as solid freeform fabrication (SFF) techniques, have been used to produce components from these materials.[1] In these methods, components are not produced by shaping in a die or mould followed by removal of material as in conventional forming routes. Instead, materials are added to build components[2] and micro-engineered structures.[3]

One of the SFF methods developed in recent years is direct ceramic ink-jet printing (DCIJP), which deploys ink-jet printers with nozzle diameters in the range of 20–200 μm to print ceramic suspensions (also referred to as inks), based on organic[1,4] or aqueous[5,6] media. Droplets of ink are delivered through the nozzle and deposited according to the pattern generated by a computer. Layer-by-layer, the part is formed.[7]

Two types of DCIJP are currently being developed. Firstly, drop-on-demand jet printers have been used to print a variety of small parts using ceramic inks which are essentially non-conducting.[8] Secondly, continuous jet printers have been used in which a stream of conductive liquid is broken into droplets, charged deflected and printed.[9] In both instances a well-dispersed ceramic suspension passes directly through the printer nozzle and production will cease rather than deposit agglomerates of a size greater than or equal to the nozzle diameter. Therefore, DCIJP has the attributes of a fail-safe manufacturing technique for ceramics.

The state of dispersion in the ceramic suspension influences strongly the viscosity of the ink and hence its flow through the printer nozzle.[10,11] In the case of DCIJP using a continuous printer, the addition of an electrolyte is necessary in order to confer conductivity to the ink. However, electrolytes can regress the state of dispersion.[11] Recent work has shown that a viscosity of $<100\,\mathrm{mPa\,s}$, a surface tension of $\sim 25\,\mathrm{mN\,m^{-1}}$ and a DC conductivity of $>100\,\mathrm{mS\,m^{-1}}$ are desirable for continuous DCIJP.[11] Figure 1 shows a resume of the criteria governing the major property requirements of ceramic inks for continuous jet printing. In

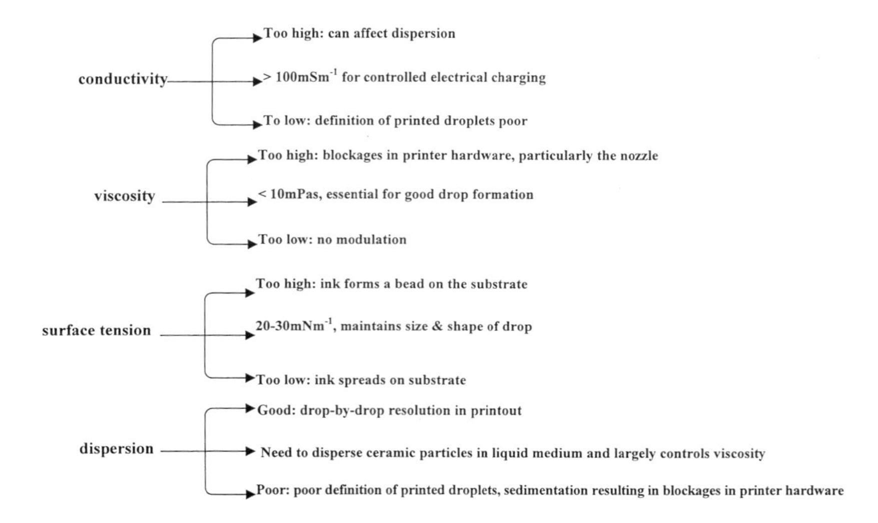

Figure 1. Summary of requirements of ceramic inks for continuous ink jet printing.

addition to these criteria, it may be possible to use the apparent pH of a ceramic ink as an overall indicator of suitability for its printing and work in this respect is in progress.[12]

It is important to increase the ceramic throughput from the printer. The use of high shear mixing using a twin roll mill was found to be effective in increasing the volume fraction of ceramic in printable suspension.[12] Dispersion of the ceramic powder in the ink can be enhanced further by ultrasonic disruption.[13,14]

In the present work, two methods to produce zirconia inks have been investigated: (i) triple roll milling and (ii) triple roll milling followed by ultrasonic disruption. The key properties of the inks, namely, sedimentation characteristics, viscosity, surface tension and conductivity, prepared using methods (i) and (ii) have been measured and printing trials have been carried out using a Linx continuous ink-jet printer.

2. EXPERIMENTAL DETAILS

2.1. *Materials*

The ceramic powder used was grade HSY3 zirconia doped with 5.4 wt% yttria having a density of 5954 kg m^{-3}, average particle size of 0.41 μm and a specific surface area of 7 m^2 g^{-1}. The powder was in the non-spray dried form and contained loose natural agglomerates. The dispersant, binder, plasticiser and solvent used in the ink formulation are given in Table 1.

2.2. *Composition and Preparation of Ceramic Inks*

The composition of the ink used was based on prior experience[11] and constituents used are described in Table 1. A binder and a plasticiser were added because of the need for mechanical strength in the dried print.

Zirconia powder, dispersant, binder and plasticiser were mixed manually with only enough addition of industrial methylated spirit (IMS) to form a paste. This mixture was subsequently triple-roll milled (Cox Machines Ltd., Bristol, UK) at ambient temperature for 15

Table 1. Material used in the formulation of ceramic inks

Components	Designation	Density kg m^{-3}	Supplier
Powder, Zirconia	HSY3	5954	Daiichi-Kigenso, Japan
Dispersant, Zephrym PD3315	ATS	1100	Uniquema, UK
Binder, Polyvinyl butyral BN18	PVB	1100	Wacker chemicals, Germany
Plasticiser, Dibutyl Sebacate	DBS	916	Aldrich Chemicals, UK
Solvent, Industrial Methylated spirit	IMS	811	Merck Ltd., UK

Table 2. Composition of the ceramic ink

Material	Vol. %
HSY3	5.00
ATS	1.36
PVB	1.82
DBS	1.82
IMS	90.00

minutes. The gap between the rolls was set at 0.5–1.0 mm to generate a high shear stress which helps to de-agglomerate the powder. During milling the material was re-distributed laterally at frequent intervals in order to enhance the dispersion of zirconia. The paste was taken off the rolls when most of the solvent had evaporated and flexible thin sheets of material were formed. Subsequently these sheets were dried in a vacuum oven at ambient temperature for 12 hours. During subsequent dilution sufficient IMS was added to attain the 5 vol.% loading of zirconia.

The inks were subsequently agitated for 2 hours to produce a suspension. 250 ml samples of ink were subjected to ultrasonic vibration using a 11 mm × 12.75 mm rectangular cross-section standard horn (Sonifier II/250, Branson Ultrasonics, Hayes, Middlesex, UK) for 15 minutes. Ultrasonic disruption was set to deliver 75 W in the ink samples kept in a conical flask surrounded by cooling water at 1°C.

2.3. *Characterisation of Zirconia Inks*

2.3.1. *Sedimentation tests*

15 ml of the inks were poured into test tubes that had been calibrated for volume against height with liquid from a burette. These were stoppered, sealed and left undisturbed for a period of 2.6 Ms (~30 days). At regular intervals during this period, the gravitational sediment volumes were measured to an accuracy of ±0.05 ml.

2.3.2. *Viscosity*

The apparent viscosity of the inks was measured at 25°C using a calibrated U-tube type

BS/IP/RF reverse flow viscometer in accordance with BI188: 1997. The accuracy of the time measurement was about ±1 s corresponding to 0.07 mPa s for the viscometer used.

2.3.3. Surface tension

The surface tension of the inks was measured using a torsion balance (White Electrical Instrument Company Ltd, Malvern Link, Worcestershire, UK) using flame-cleaned glass slides. Distilled water having a surface tension of 72 mN m^{-1} was used to calibrate the instrument.

2.3.4. DC conductivity

The DC conductivity of the inks was measured at ambient temperature using a conductivity meter (HI 9033, Hanna Instruments Ltd, Leighton Buzzard, Bedfordshire, UK). The meter was calibrated using a standard solution of KCl supplied by Hanna Instruments Ltd.

2.3.5. Scanning Electron Microscopy

The state of the dispersion in the inks was studied using a JEOL 6300 scanning electron microscope. Using a pipette, a drop of ink was transferred to an aluminium stub and allowed to spread. When dry, the samples were sputter coated with gold for 10 minutes.

2.3.6. Printing

A Linx 6200S continuous ink-jet printer was used. This printer has three main sections:

(i) the ink control unit which contains the ceramic ink,
(ii) the printhead that creates, directs and prints droplets of ink, and
(iii) the sliding table fitted with an optical track providing registration for automatic multilayer printing.

Samples of zirconia ink both with and without being subjected to ultrasonic disruption were prepared for printing. NH_4NO_3 (density 1720 kg m^{-3} supplied by Merck Ltd, UK) was used as the electrolyte and 1.2 g of it was added to each 100 ml of ink. The inks were filtered prior to printing by using polycap disposable filtration capsules of 10 and 5 μm (Arbor Technologies Inc., Ann Arbor, USA) with a micropump (Micropump Inc., Vancouver, USA).

The filtered zirconia ink was supplied under pressure to an ink gun and forced out through a small nozzle, 62 μm in diameter. As the ink passed through the nozzle, it was piezo-electrically pulsed and modulated. The stream of ink broke up into a continuous series of drops which were equally spaced and were of the same size. A charge electrode surrounds the jet at the point where the drops separate from the ink stream. A voltage was applied between the charge electrode and the drop stream. When the drop breaks off from the stream it carries a charge proportional to the applied voltage at the instant at which it breaks off. By varying the charge electrode voltage (50–300 V) at the same rate as the drops are produced, it was possible to charge every drop to a predetermined value.

The drop stream continues its flight and passes between two deflector plates, which were maintained at a constant potential, typically ±3 kV. In the presence of this field, drops were deflected towards one of the plates by an amount proportional to the charge carried. Drops of ink which were uncharged, were undeflected and were collected by a gutter to be recycled into the ink reservoir. The drops which were charged and hence deflected, deposited on the substrate attached to the sliding table.

The droplets were visible through a window at the print-head with the aid of LED illumination. Controlling the modulation frequency changed the shape of the droplets. Ink viscosity can impose a minor change in the mass of the drop and this affects the final printing position of the drop. However, precise placement of drops is dependent on the speed of drops as they travel through the deflector plates and for this reason, the time of flight of the drops was constantly monitored and compared with the optimum value printed on the print-head conduit.[15] A 'feedback-loop' was used to increase or decrease the ink pressure, thus maintaining the correct speed of drops for accurate placement.

3. RESULTS AND DISCUSSION

Sedimentation behaviour of a suspension is often useful in assessing the dispersability of ceramic powder and gives an indication of the degree of deflocculation.[16] Finely dispersed particles tend to remain suspended in the liquid while undispersed agglomerates or flocculated structures settle under gravity at a much higher rate.

Zirconia particles in the ink made using only triple roll milling began to settle almost instantaneously but settling in the ink made by triple roll milling followed by ultrasonic disruption was slower and there was no clear interface between the supernatant and the suspension. After about one day, in this ink, a milk-like liquid (supernatant) and a denser liquid (suspension) were observed at the top and bottom of the test tube, respectively. Variation of sediment volume with time for the inks which were prepared by triple-roll milling (ink A) and triple-roll milling followed by ultrasonic disruption (ink B) are presented in Fig. 2. Ink A shows a sediment volume almost instantaneously. In contrast, ink B did not show any

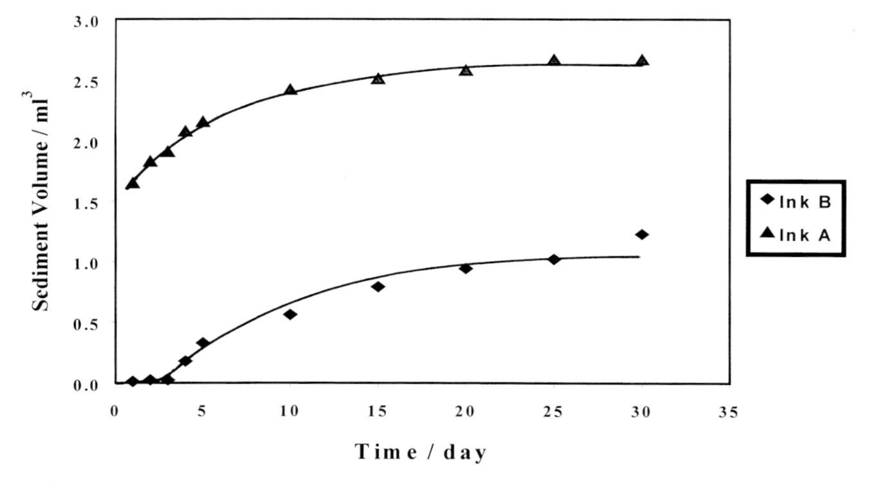

Figure 2. Variation of sediment volume of the inks prepared by triple roll milling (ink A) and triple roll milling followed by ultrasonic disruption (ink B).

sedimentation for three days. The final sediment volume of ink A was more than twice that of ink B. Further improvement in the dispersion of ink B is sought by the use of a high-energy bead mill and preliminary work shows that a zirconia ink which does not show any sedimentation for four days can be produced.[17]

Scanning electron micrographs of the ink prepared by triple-roll milling clearly shows the presence of agglomerates (Figure 3a). Ultrasonic disruption helps to break down some of

(a)

(b)

Figure 3. Scanning electron micrographs of (a) ink A and (b) ink B.

Figure 4. Printed pattern (single layer) obtained using ink B with a 2.5 vol.% zirconia (after further sedimentation for 16 hours).

these agglomerates (Figure 3b). The intensity of the ultrasonic vibration is sufficient to reduce the absolute pressure creating a large number of very small bubbles. The collapse of these bubbles during the ultrasonic cycle produces shock waves of sufficient magnitude to disrupt the agglomerates.

Ceramic inks should contain sufficient powder for rapid deposition and yet have appropriate viscosity, surface tension and in the case of continuous printing, have sufficient conductivity. The lack of conductivity results in printing poorly defined characters, but sufficient conductivity does not guarantee good quality printing if the dispersion of the ceramic in the ink is inadequate.[11] Thus, it is appropriate to remove the undispersed agglomerates by sedimentation prior to printing and this improves substantially the quality of printing.[11]

Both inks A and B had an apparent viscosity of 4 mPa s, a surface tension of 23 mN m^{-1} and a conductivity of 236 mS m^{-1}. However, the quality of printing achieved from both inks showed poor definition of characters. In the case of ink A, sedimentation over a period of 10 hours resulted in a suspension containing 1.6 vol.% zirconia having a viscosity of 2.7 mPa s. There is little point in printing this suspension as the ceramic throughput will be very small. A similar treatment of ink B resulted in a suspension containing 2.5 vol.% zirconia and its viscosity was 3 mPa s. It produced a print with well-defined characters (Figure 4) and illustrates the effect of ultrasonic disruption in producing better dispersion. In fact, ultrasonic disruption has led to the increase of the zirconia volume in printable ink to 4.4 vol.%.[18]

CONCLUSIONS

Ultrasonic disruption has demonstrated short-term effectiveness in improving the dispersion of zirconia in the ink prepared for direct ceramic jet printing. The initial sedimentation rate was markedly slower compared with the ink prepared with triple roll milling only. However, undispersed powder still remained in the ink and it could only be printed successfully after further sedimentation. Eventually, an ink containing 2.5 vol.% zirconia was put through the continuous ink-jet printer and a ceramic pattern with well-defined characters was printed.

ACKNOWLEDGEMENTS

The authors wish to thank EPSRC for funding this research under Grant GR/M07731 and Linx Printing Technologies plc, St. Ives, Cambridgeshire, UK for supporting ink-jet printing research at Queen Mary & Westfield College. B. Y. Tay wishes to thank Gintic Institute of Manufacturing Technology, Singapore for funding her PhD research on ceramic ink-jet printing.

REFERENCES

1. M. J. Edirisinghe: in *Processing and Fabrication of Advanced Materials VII*, T. S. Srivatsan and K. A. Khor (Editors), The Minerals, Metals & Mater. Soc. (USA), 1998, 139–150.
2. J. G. Conley and H. L. Marcus: *J. Manuf. Sci. Trans. ASME*, 1997,, **119**, 811–816.
3, M. Mott and J. R. G. Evans: *Mater. Sci. Eng.*, 1999, **271**, 344–352.
4. W. D. Teng and M. J. Edirisinghe: *British Ceramic Trans.*, 1998, **97**, 169–173.
5. C. E. Slade and J. R. G. Evans: *J. Mater. Sci. Lett.*, 1998, **17**, 1669–1671.
6. J. Windle and B. Derby: *J. Mater. Sci. Lett.*, 1998, **18**, 87–90.
7. M. Mott, J. H. Song and J. R. G. Evans: *J. Amer. Ceram. Soc.*, 1999, **82**, 1653–1658.
8. Q. F. Xiang, J. R. G. Evans, M. J. Edirisinghe and P. F. Blazdell: *Proc. Inst. Mech. Eng. (UK)*, 1997, **B211**, 211–214.
9. P. F. Blazdell, J. R. G. Evans, M. J. Edirisinghe, P. Shaw and M. J. Binstead: *J. Mater. Sci. Lett.*, 1995, **14**, 1562–1565.
10. W. D. Teng, M. J. Edirisinghe and J. R. G. Evans: *J. Amer. Ceram. Soc.*, 1997, **80**, 486–494.
11. W. D. Teng and M. J. Edirisinghe: *J. Amer. Ceram. Soc.*, 1998, **81**, 1033–1036.
12. H. Rashid, M. J. Edirisinghe and J. R. G. Evans: to be published.
13. H. Rashid, B. Y. Tay and M. J. Edirisinghe: *J. Mater. Sci. Lett.*, 2000, **19**, 799–801.
14. B. Y. Tay, H. Rashid and M. J. Edirisinghe: *J. Mater. Sci. Lett.*, in press.
15. Linx 6200 Series User Manual, Linx Printing Technologies plc, St Ives, Cambridgeshire, UK, page 171.
16. T. Chartier, S. Souchard, J. F. Baumard and H. Vesteghem: *J. Euro. Ceram. Soc.*, 1996, **16**, 1283–1291.
17. H. Rashid, M. J. Edirisinghe and J. R. G. Evans: to be published.
18. J. H. Song, M. J. Edirisinghe and J. R. G. Evans: *J. Amer. Ceram. Soc.*, 1999, **82**, 3374–3380.

In situ Monitoring of Ceramic Sintering Using Impedance Spectroscopy

X. WANG and P. XIAO*
Department of Mechanical Engineering, Brunel University, Uxbridge, UB8 3PH, UK

ABSTRACT

Impedance measurements were made *in situ* while a clay compact was being fired at different temperatures. The measured impedance spectra consist of a high frequency (HF) semicircular arc and a low frequency (LF) tail. By employing an equivalent circuit of the clay compact to simulate the impedance spectra, values for resistances and the parameters of constant phase elements (CPEs) corresponding to both bulk specimen and electrode effects have been obtained. It was found that impedance spectra were sensitive to the formation of the amorphous liquid phase during heating of the green compact. The change in impedance spectra during isothermal sintering indicates densification of the clay compact. Impedance spectroscopy has been found to be a promising non-destructive tool for monitoring the real time sintering process in ceramics.

1. INTRODUCTION

In situ monitoring of ceramic sintering is extremely important for understanding the sintering mechanisms and improving the efficiency of ceramic manufacturing. Impedance spectroscopy (IS) is a relatively mature, cheap and simple technique for non-destructive testing, which has been widely used to characterise the electrical properties of the materials and relate the changes in these electrical properties to microstructural changes occurring in the materials. Recently, growing interest has been developed in employing IS to characterise the microstructure of ceramic materials,[1–3] particularly for studying microstructure evolution in cement during hydration and hardening.[4–6]

In impedance measurements, a sinusoidal potential variation is applied to the test electrodes sandwiching a specimen. Impedance $Z(= V/I)$ is obtained by measuring the magnitude and phase shift of the resulting current over a range of frequencies of an AC power supply. Impedance diagrams are normally expressed as Nyquist and Bode plots. The Nyquist plot shows the imaginary part of the impedance versus the real part of the impedance. The Bode plot shows the impedance as a function of frequency and phase angle.

The purpose of this work is to use impedance measurements as a non-destructive method for monitoring *in situ* the sintering process of clay-based ceramics. Clay-based ceramics have a wide range of applications and they are the most complex ceramic materials.[7] Liquid phase sintering is the dominant mechanism in the sintering of clay based ceramics.[8] The formation of liquid phase is a crucial step in the process as it initiates compact shrinkage. The quantity and the viscosity of the liquid phase are two important factors that determine the kinetics of sintering.[8, 9] Liquid phases formed in ceramic compacts at high temperature are ionic conductors and thus act as electrolytes.[10] The microstructural parameters, such as the quantity and chemical compostion of liquid phase, the porosity content and the distribution of different phases will affect the sintering process. Further, these parameters keep changing

* Corresponding author.

as sintering proceeds. Thus, monitoring sintering of clay ceramics using impedance spectroscopy will provide important information on sintering phenomena in clays.

2. EXPERIMENTAL PROCEDURE

2.1. *High Temperature Measurement Rig*

A rig for impedance measurements at high temperature was built as shown in Fig. 1. In the rig, a wire spring was used to apply a constant pressure of about 5 N to the specimen to ensure an intimate contact between the electrodes and the specimen. The electrodes were two platinum foils bonded to Al_2O_3 plates and connected to the electrodes of the impedance analyser. Since the resistance of the leads connecting the electrodes to the impedance analyser is very small, when compared with the tested samples, the influence of the leads on impedance measurement is negligible.

2.2. *Sample Preparation*

Ball clay (Dorset, HYMOD PRE) was dry ball-milled for 5 h to produce a powder of uniform particle size. The chemical composition and particle size distribution specified by the powder provider are given in Tables 1 and 2 respectively. Pellets 13 mm in diameter and 2~5 mm in thickness were prepared by uniaxial pressing using a pressure of 120 MPa.

2.3. *Impedance Measurements*

A Solartron SI 1255 HF frequency response analyser coupled with a 1296 dielectric interface (Solartron, UK) was used for impedance measurement. Data acquisition was

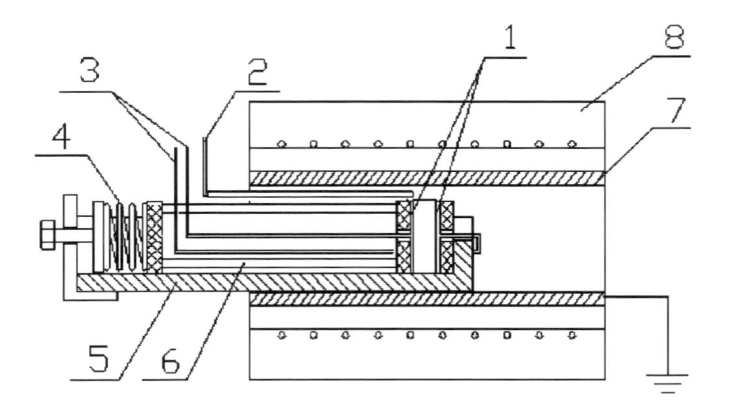

1. ELECTRODES 2. THERMAL COUPLE 3. LEADS
4. SPRING 5. CERAMIC HOLDER 6. SILICA TUBE
7. SHIELDING 8. TUBULAR FURNACE

Figure 1. Schematic illustration of the rig for high temperature impedance measurement.

Table 1. Chemical composition of the ball clay (wt%).

SiO_2	Al_2O_3	Fe_2O_3	TiO_2	CaO	MgO	K_2O	Na_2O	L.O.I	C
54	30	1.4	1.3	0.3	0.4	3.1	0.5	8.8	0.3

Table 2. Particle size distribution of the ball clay (wt %).

<5 μm	<2 μm	<1 μm	<0.5 μm
96	88	79	67

undertaken using a PC computer in real time while the clay compact was being fired at different temperatures. Measurements were taken over a frequency range of 10^6 to 0.1 Hz using 1 V applied voltage with 3~6 readings per decade of frequency. Measurements with voltages lower than 0.2 V resulted in unstable responses due to the high impedance of the sample, while measurements with voltages of 0.2~3 V gave stable signals and reproducible results. Therefore, 1 V was chosen for all impedance measurements. Measurements at a frequency of 10^6 Hz were made to determine the dielectric loss at different temperatures. Impedance spectra were analysed using the software package 'Zview for Windows' (Scribner Associates).

3. RESULTS AND DISCUSSION

3.1. *The IS spectra*

Isothermal impedance measurements were made at different temperatures during firing of the clay compact. The impedance spectra measured at temperatures above 500°C are reproducible. At temperatures below 500°C, several factors could affect impedance measurements, e.g. incomplete contact between sample and electrode, a short circuit through a less resistive path in the clay, and the presence of moisture in clay.

Figure 2 shows the Nyquist plots from impedance measurements at 700, 800, 900, and 1000°C. There is a high frequency (HF) semicircular arc and a low frequency (LF) tail in each of the Nyquist plots. Both the HF arc and the LF tail become smaller with increasing temperature and the cut-off frequency between the two increases from 4.6 Hz at 700°C to 4600 Hz at 1000°C. Measurements made using samples with varying thickness showed that the thickness of the sample affected the HF arc radii and cut-off frequency in the Nyquist plot, but had little effect on the LF tail. Thus the HF arc corresponds to the response of the clay, whereas the LF tail corresponds to the electrode effect, which relates to the polarisation of the electrode. This phenomenon is similar to that found in studies of cement paste using IS.[4–6]

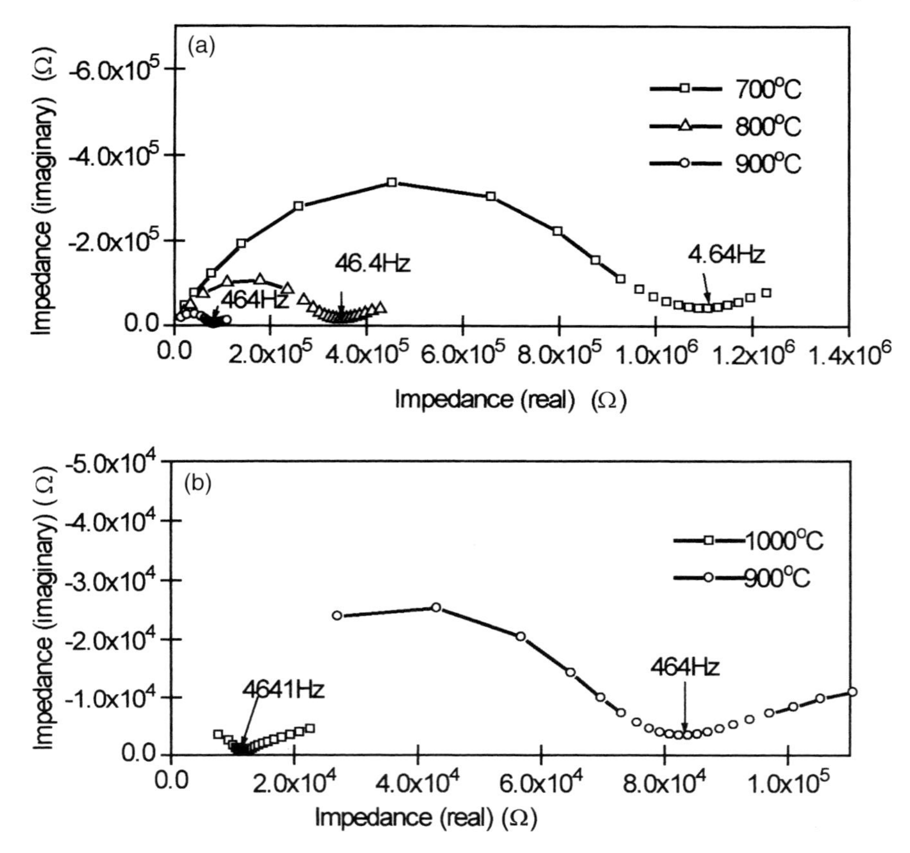

Figure 2. (a) Nyquist plots of the clay compact at 600, 700 and 800°C, (b) Nyquist plots of the clay compact at 900 and 1000°C.

3.2. Simulation of impedance spectra using equivalent circuit model

The equivalent circuit (Fig. 3) consisting of resistor (R) and constant phase elements (CPEs) was used to simulate measured impedance spectra. Here CPEs were used to represent the electrical response of complex materials with a range of relaxation frequencies. The impedance of the CPE is given by

$$Z_{\text{CPE}}(j\omega) = A^{-1}(j\omega)^{-n} \tag{1}$$

where A is a constant that is independent of frequency, ω is angular frequency and $j = -1^{1/2}$. When $n = 1$, the CPE represents an ideal capacitor; when $n = 0$, the CPE acts as a pure resistor.[11] Thus, a CPE can represent a wide variety of non-ideal elements.

This equivalent circuit model gave very good fitting to the impedance spectra. The results from the fitting are summarised in Table 3, where it can be seen that resistance decreases with increasing temperature while the parameters for the CPEs increase with increasing temperature. A_e (for electrode) is about five orders of magnitude larger than A_b (for a bulk

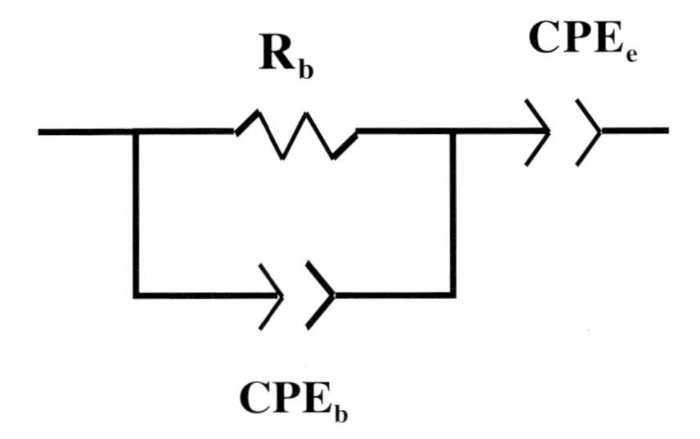

Figure 3. Equivalent circuit of a clay pellet for simulation of impedance spectra.

Table 3. Results of simulation using an equivalent circuit.

Temperature (°C)	R_b (ohm)	A_b ($ohm^{1-n}F^{-n}$)	n_b	A_e ($ohm^{1-n}F^{-n}$)	n_e
700	1.04×10^6	1.99×10^{-10}	0.730	...	...
800	3.20×10^5	2.70×10^{-10}	0.730	...	...
900	7.60×10^4	3.57×10^{-10}	0.740	2.55×10^{-5}	0.157
1000	7.83×10^3	4.50×10^{-10}	0.746	7.79×10^{-5}	0.250
1100	2.20×10^3	9.90×10^{-10}	0.768	1.04×10^{-4}	0.274

specimen), but n_b (CPE exponential index for a bulk specimen) is much closer to 1 than n_e (CPE exponential index of the electrodes). This indicated that bulk specimen acts more like a capacitor whereas the electrode acts more like a resistor.

3.3. *Sensitivity of the IS to the formation of liquid phase*

Figure 2b shows a small part of the semicircle corresponding to the bulk specimen when the impedance spectra were measured at temperatures below 1000°C. The relaxation frequency was well above 1 MHz. This suggests that the bulk clay behaved like a conducting electrolyte at 1000°C, which may be due to the formation of a continuous conductive liquid phase in the clay compact. In the Arrhenius plot showing the relationship between conductance and temperature (Fig. 4), there is a change in gradient at about 920°C. This indicates that the conducting mechanism in the clay changes when the temperature exceeds 920°.

The change in dimensions of the bulk specimen during firing was measured using a dilatometer. Figure 5 shows that significant shrinkage occurs only at temperatures above 950°C. This confirms that considerable amounts of liquid phase form at 950°C. As the dominant mechanism in the sintering of clay-based ceramics is liquid phase sintering, significant shrinkage could not occur unless a certain amount of liquid phase was present.[8–9]

In a previous study, electrical conductivity measurements were made to detect a liquid phase suspected to form in the Al_2O_3–1 mol% TiO_2–0.5 mol% $NaO_{1/2}$ system.[10] It was found

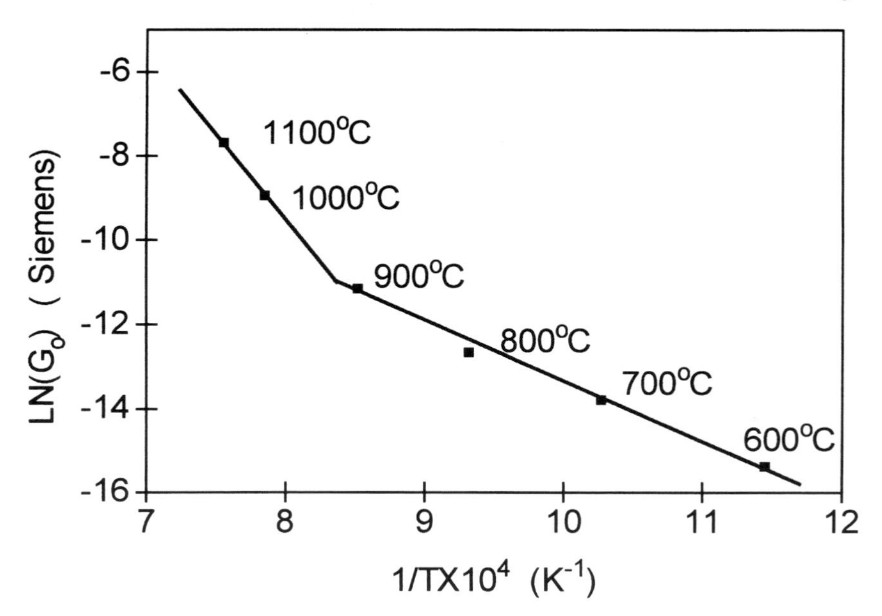

Figure 4. Arrhenius plot of bulk conductance as function of temperature.

that the dc electrical measurements were sensitive enough to indicate the presence of a small amount of liquid phase. Impedance spectroscopy is a much more sophisticated technique than the conductivity measurement technique. More information, e.g. capacitative effects,

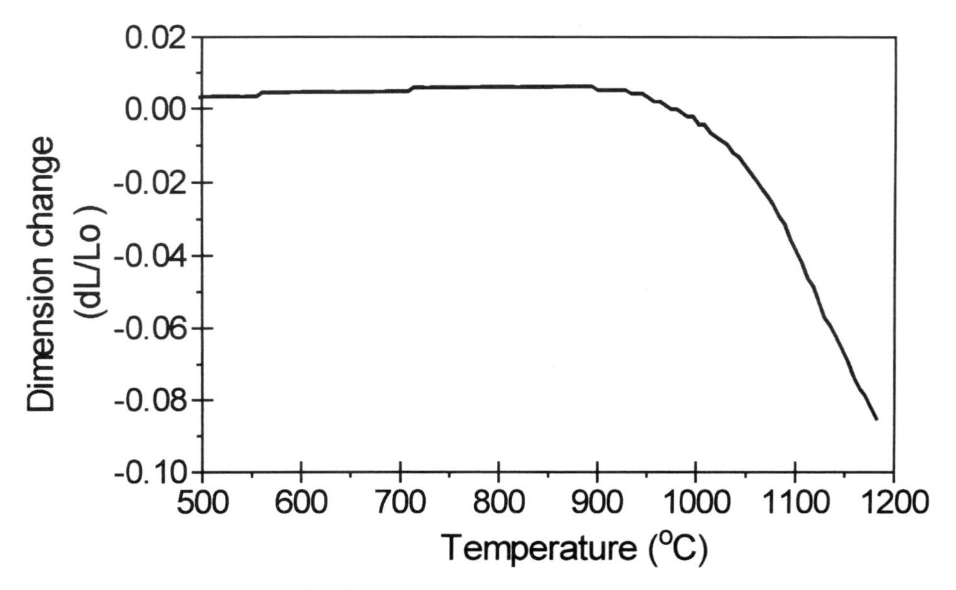

Figure 5. Shrinkage curve of a clay compact measured using a dilatometer, the heating rate was $5\,\mathrm{K\,min^{-1}}$.

dielectric loss and electrode polarisation, can be obtained from IS spectra. Thus impedance spectroscopy is a potentially more powerful tool for studying sintering mechanisms.

4. MONITORING DENSIFICATION PROCESS

To monitor the densification process of clay, impedance measurements at different time intervals have been made at various temperatures. Little change in spectra measured at 800°C (Fig. 6a) was found and the spectra measured at different periods overlap completely. In contrast, the spectra measured at temperatures above 950°C (Fig. 6b~d) showed a gradual shift towards the left. This is due to the increasing density of the clay during sintering. Meanwhile the other parameters relating to the spectrum, i.e. the dielectric loss, the cut-off frequency and the length of LF tail etc., remained almost unchanged. This means that the shrinkage affected the bulk resistance only, while the other parameters relating to the spectrum appeared to be affected mainly by the temperature.

Figures 6b–d show the multiple impedance spectra of the clay at 1050, 1100 and 1150°C respectively. The shifts in the spectra are more marked at the beginning of sintering and at the higher temperatures. The shifts are smaller at the later stage of sinterings. This indicates that at high temperature, the densification occurs more quickly at the beginning and then slows down as shown in a previous study.[9] Further work is needed to establish a relationship between the relative density and the impedance parameters of the clay.

Since each measurement in the impedance spectra was obtained over 1~4 min, this does not correspond to a stable state of the clay, especially at higher temperatures when the compacts were undergoing rapid densification. However, there was little change in the features

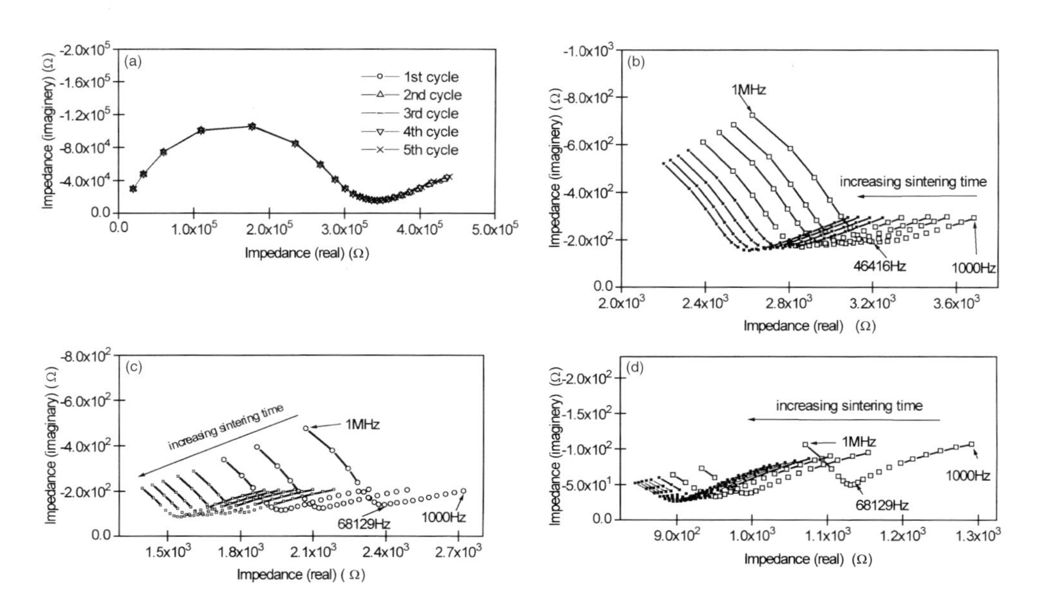

Figure 6. Multiple impedance measurements at time intervals of 5 min, each measurement taking 2.62 min; at (a) 800°C, (b) 1050°C, (c) 1100°C and (d) 1150°C.

of impedance spectra during the sintering process. Thus, multiple impedance spectroscopy is still a powerful method for monitoring the densification process of clay-based materials.

4 SUMMARY AND CONCLUSION

Impedance spectroscopy (IS) is an effective technique to examine the behaviour of a ceramic green compact during firing. In the Nyquist plot of IS, the bulk specimen effect and the electrode effect are well separated. Impedance measurements can be used to determine the formation of liquid phase in a clay during sintering.

Shrinkage of the clay compact leads to a shift in the impedance spectra during sintering, but has no influence on the electrode contribution. The shift of the spectra is an indication of the densification process. Hence, impedance spectroscopy can be used to monitor *in situ* the ceramic sintering process.

REFERENCES

1. E. N. S. Muccillo and M. Kleitz: 'Impedance spectroscopy of Mg-partially stabilised zirconia and cubic phase decomposition', *J. Eur. Ceram. Soc.*, 1996, **16**, 453–465.
2. C. M. S. Rodrigues, J. A. Labrincha and F. M. B. Marques: 'Monitoring of the corrosion of YSZ by impedance spectroscopy', *J. Eur. Ceram. Soc.*, 1998, **18**, 95–104.
3. M. C. Steil and F. Thevenot and M. Kleitz: 'Densification of yttria-stabilised zirconia-impedance spectroscopy analysis', *J. Electrochem. Soc.*, 1997, **1**, 390–398.
4. W. J. McCarter, S. Garvin and N. Bouzid: 'Impedance measurement on cement paste', *J. Mater. Sci. Lett.* 1988, 1056–1057.
5. B. J. Christensen, T. O. Mason and H. M. Jennings: 'Influence of silica fume on the early hydration of portland cements using impedance spectroscopy', *J. Am. Ceram. Soc.*, 1992, **75**, 939–945.
6. B. J. Christensen, R. T. Coverdale, R. A. Olseon, S. J. Ford, E. J. Carboczi, H. M. Jennings and T. O. Mason: 'Impedance spectroscopy of hydrating cement-based materials: measurement, interpretation, and application', *J. Am. Ceram. Soc.*, 1994, **77**, 2789–2804.
7. W. M. Carty and U. Senapati: 'Porcelain-raw materials, processing, phase evaluations, and mechanical behaviour', *J. Am. Ceram. Soc.*, 1998, **81**, 3–20.
8. W. D. Kingery, H. K. Bowen and D. R. Uhlmann: 'Grain growth, sintering and vitrification', in *Introduction to Ceramics*, edited by E. Burke, B. Chalmers and J. A. Krumhansl, 2nd edn, John Wiley & Sons, New York, NY, 1976, 448–575.
9. M. R. Anseau, M. Deletter and F. Cambier: 'The separation of sintering mechanisms for clay-based ceramics', *Trans. J. Br. Ceram. Soc.*, 1981, **80**, 142–146.
10. P. E. D. Morgan and M. S. Koutsoutis: 'Electrical conductivity measurements to detect suspected liquid phase in the Al_2O_3-1mol% TiO_2-0.5 mol% $NaO_{1/2}$ and other systems', *J. Am. Ceram. Soc.*, 1986, **67**, c254–c255.
11. S. T. Amaral and I. L. Muller: 'Effect of silicate on passive films anodically formed on iron in alkaline solution as studied by electrochemical impedance spectroscopy', *Corrosion*, 1999, **55**, 19–23.

Fabrication of $La_{0.8}Sr_{0.2}MnO_3$/Metal Interfaces with Low Electrical Resistance

J. Q. LI and P. XIAO*

Materials Engineering Group, Department of Mechanical Engineering, Brunel University, Uxbridge, UB8 3PH UK

ABSTRACT

In order to develop metal/cathode connections for the construction of solid oxide fuel cells (SOFCs), interfaces between $La_{0.8}Sr_{0.2}MnO_3$ (LSMO) and metals have been fabricated using screen printing, followed by sintering at 1200°C. The LSMO/Fecralloy interface fabricated in both air and vacuum showed a high electrical resistance at temperatures up to 800°C. Both alumina and mixed oxide $M_2O_3.nAl_2O_3$ were formed at the interface during the fabrication processes and following heat treatments. To reduce the electrical resistance of the interface, the LSMO/Ducrolloy (Cr–5Fe–Y_2O_3) interface was fabricated in air, in which a Cr_2O_3 layer was formed at the interface and a spinel phase $(Mn, Cr)_3O_4$ was formed inside the porous LSMO. The electrical resistances of both phases are high at room temperature, but become negligible at temperatures above 500°C. However, the evaporation of CrO_3 in high temperature environments increases the electrical resistance of the LSMO layer. It is necessary to develop a barrier coating on the Cr–5Fe–Y_2O_3 substrate for the prevention of CrO_3 evaporation.

1. INTRODUCTION

Solid oxide fuel cells (SOFCs) are electrochemical power sources that directly convert the energy of a chemical reaction into electrical energy with high efficiencies.[1] They have the potential for an environment friendly supply of energy. SOFCs typically consist of electrolyte, cathode, anode and interconnector. Considering the chemical stability, thermal expansion match and electrical conductivity, at present, yittria stablised zirconia (YSZ), YSZ/NiO, Sr-doped lanthanum manganite (e.g. Sr-doped $LaMnO_3$) and Sr-doped $LaCrO_3$ are commonly used as electrolyte, anode, cathode and interconnector in SOFCs respectively.[1–3]

The Sr-doped $LaCrO_3$ interconnect material has very low mechanical strength and poor machinability when compared with metals. Substitution of this interconnect material by metals has the advantage of a higher electronic conductivity, higher ductility, higher heat conductivity and better workability. The interconnect materials must be chemically stable in both reducing and oxidising environments at high temperature, have high thermal conductivity, gas tightness and matching thermal expansion with the electrolyte (YSZ, $11 \times 10^{-6}\ K^{-1}$)[4] and cathode (Sr-doped $LaMnO_3$ (LSM), $12 \times 10^{-6}\ K^{-1}$).[5] Rak *et al.*[6] suggested that a $LaCrO_3$/Cr composite could be one metallic interconnector candidate because $LaCrO_3$ possesses electronic conductivity while Cr metal has relatively low thermal expansion. A few studies have been carried out to investigate the possibility of using Cr-containing alloys, such as Fecralloy and Ducrolloy, as interconnectors in SOFCs.[4, 7–10] These studies have been focused on studying the oxidation resistance and microstructures of Cr containing

* Corresponding author.

alloy/LSMO interfaces. However, little research has been done to investigate the electrical properties of the interface in relation to the microstructure of the interface, which is important for application of the alloys as interconnectors and LSMO as cathodes in SOFCs. In this work, interfaces were fabricated between $La_{0.8}Sr_{0.2}MnO_3$ and Fecralloy (Fe–22Cr–4.8Al–0.3Si–0.3Y in wt %) or Ducrolloy (Cr–5Fe–1Y_2O_3 in wt %) by screen printing and sintering at 1200°C for 2 h in air. The electrical properties of different interfaces have been examined using impedance spectroscopy. The effects of interfacial microstructure on the electrical properties of the interfaces has been investigated.

2. EXPERIMENTAL

Two types of metallic materials were used to fabricate LSMO/metal interfaces: Fecralloy and Ducrolloy. Fecralloy foils with the thickness of 1.0 mm were obtained from Goodfellow Ltd. UK. The Ducrolloy alloy was prepared using a melting technique. The Sr-doped lanthanum manganite, $La_{0.8}Sr_{0.2}MnO_3$ powder (99.9%, 2 μm), was supplied by Pi-Kem Ltd. UK.

Both Fecralloy and Ducrolloy were polished using the 1200 grit SiC paper to remove the oxide layer at the surface and ultrasonically cleaned in acetone. The LSMO powder was mixed with Blythe binder (John Mathew Ltd. UK) containing 50% solvent terpineol and 50% ethyl cellulose to make a paste. The paste was then screen printed onto the Fecralloy foil or the Ducrolloy foil to form a layer with thickness of about 60~80 μm. The sample was initially dried at 120°C for 6 h in air. Finally, it was heated up to 1200°C in air, argon or vacuum of 5×10^{-5} torr at a heating rate of $5\,K\,min^{-1}$ and held for 2 h before cooling down to room temperature at the same rate. Some samples were post annealed at 1200°C in air for a certain period to study the stability of the LSMO/metal interface.

Cross section of the samples were examined using a scanning electron microscope (SEM) coupled with energy dispersive X-ray analysis (EDX) (Jeol JXA-840). The phases present in the thick film layer were identified using X-ray diffraction analysis (Phillips PW 1140). AC impedance of the LSMO/metal interfaces were measured using a computer-controlled Solartron SI 1255 Hf frequency response/1296 dielectric interface analyser at frequency range of 10^6 to 10^{-3} Hz. The sample was kept at the measurement temperature for half an hour before impedance measurements were taken. Zview impedance analysis software was used to analyse the impedance spectra. The metal side of the sample was polished using 1200 grit SiC paper to remove the oxide layer formed during fabrication and then cleaned with acetone. Pt foils were used as electrodes for impedance measurements. The contact resistance between the Pt and the surface of samples was found to be negligible.

3. RESULTS AND DISCUSSION

3.1. $La_{0.8}Sr_{0.2}MnO_3$ / Fecralloy Interface

Figure 1a shows a SEM micrograph of a cross-section of the LSMO/Fecralloy sample fabricated at 1200°C in air. Although a strong bond was achieved between LSMO and Fecralloy, an oxide layer of about 3.0 μm thickness was formed at the interface.

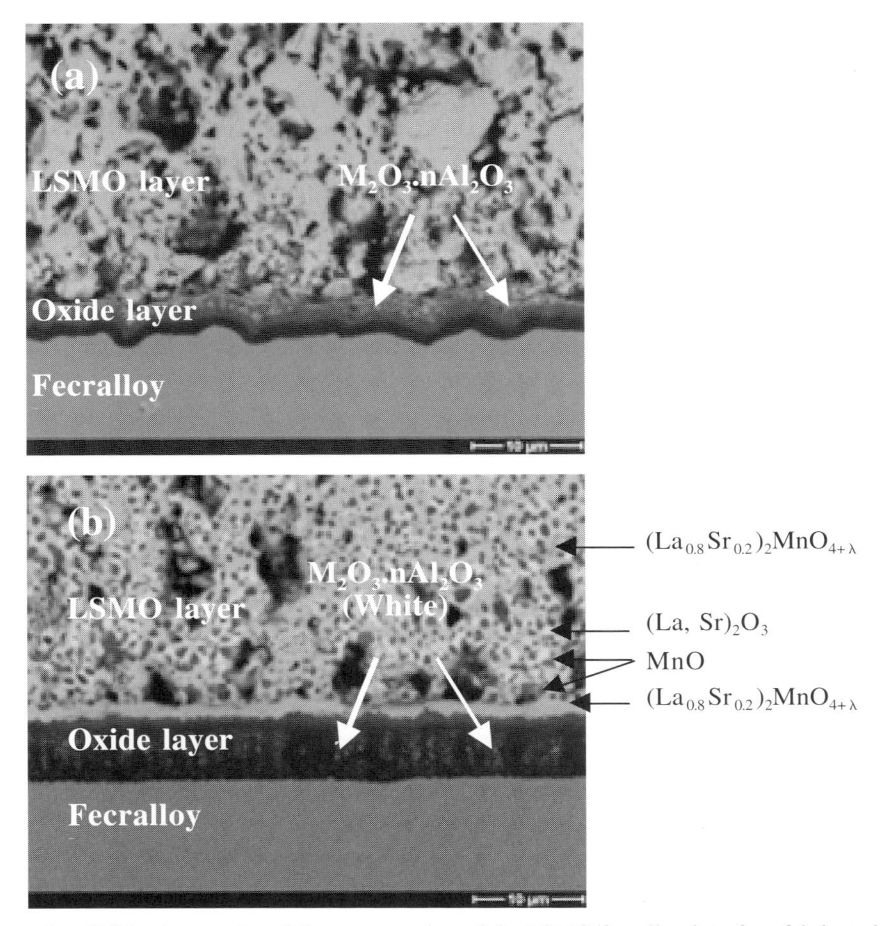

Figure 1. SEM micrographs of the cross-section of the LSMO/fecralloy interface fabricated at 1200°C for 2 h in (a) air and (b) flowing Ar; the phases in the LSMO layer in (b) are $(La_{0.8}Sr_{0.2})_2MnO_{4+\lambda}$ (light grey region), $(La, Sr)_2O_3$ (white region) and MnO (dark region).

Microanalysis indicated that the dark region of the oxide layer close to the fecralloy substrate was a high purity Al_2O_3 while the grey region close to the LSMO contains a second phase in the Al_2O_3 matrix. The second phase was identified as a mixed oxide $M_2O_3.nAl_2O_3$, where M is mainly La and Sr, together with some Mn. There is no evidence that any element from the fecralloy diffuses into the LSMO layer. X-ray diffraction analysis of the LSMO shows that only the single conducting phase $La_{0.8}Sr_{0.2}MnO_3$ exists in the LSMO layer (see Fig. 2), but the interface resistance of the sample was found to be very high due to the formation of an insulating alumina layer at the interface.

Reduction of the thickness of the interfacial oxide layer was attempted by fabricating the samples in flowing Ar or vacuum. The interface microstructures of these two samples were found to be very similar. However, it was surprising to find that the oxide layers at the interfaces of these two samples were much thicker than those formed in air. Fig. 1b shows the SEM micrograph of the cross-section of the LSMO/fecralloy interface fabricated in flowing

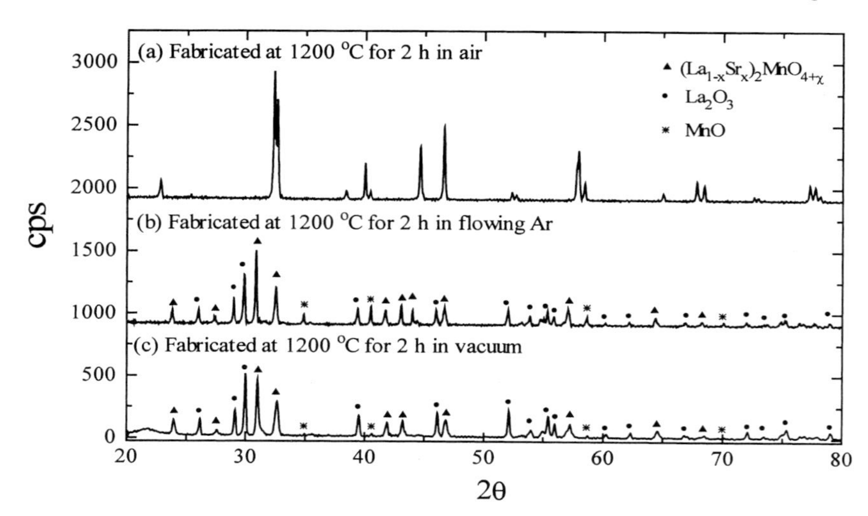

Figure 2. X-ray diffraction analysis of LSMO layer in a LSMO/fecralloy sample fabricated (a) in air, (b) in flowing pure Ar and (c) in vacuum.

Ar. The thick oxide layer is mainly Al_2O_3 phase together with some $M_2O_3.nAl_2O_3$ phase (white region). Moreover, the LSMO phase was found to decompose completely into three phases, a light grey phase, a dark grey phase and a white phase, during the sintering process. The EDX and X-ray diffraction (Fig. 2b and c) analysis of the LSMO layer indicated that the light grey phase, the dark grey phase and the white phase are $(La_{0.8}Sr_{0.2})_2MnO_{4+\lambda}$, MnO and $(La, Sr)_2O_3$ (the solution of Sr in La_2O_3) phases respectively. The light grey layer formed on the oxide layer is $(La_{0.8}Sr_{0.2})_2MnO_{4+\lambda}$. According to the phase diagram of the La–Sr–Mn–O system,[11] the compound $(La_{0.8}Sr_{0.2})MnO_3$ decomposes into $(La_{0.8}Sr_{0.2})_2MnO_{4+\lambda}$, $(La, Sr)_2O_3$ and MnO oxides at 1100°C with the equilibrium oxygen pressure of $Po_2 = 10^{-13}$ atm (10^{-8} Pa). In this work, the oxygen pressure in the vacuum furnace is about 10^{-4} Pa and higher in the flowing Ar furnace. However, it was expected that the oxygen pressure at the interface of the samples was significantly lower than the oxygen pressure in the furnace chamber.

Experiments showed that there was no decomposition of $(La_{0.8}Sr_{0.2})MnO_3$ when a pure ceramic specimen was treated using the same thermal process used for the fabrication of the LSMO/Fecralloy interface. Therefore, the LSMO/fecralloy interfacial reaction reduced the O_2 partial pressure at the interface and induced the decomposition of the $(La_{0.8}Sr_{0.2})MnO_3$ phase. The decomposition caused the poor electrical conductivity in the LSMO layer. However, it can be partially reversed after thermal treatment of the LSMO/Fecralloy interface at 1200°C for 5 h in air (Fig. 3).

Impedance spectroscopy was used to determine the electrical resistance of the LSMO/Fecralloy interfaces. These interfaces were fabricated in air or flowing Ar. The interface fabricated in flowing Ar was treated thermally in air at 1200°C for 5 h after fabrication. Figure 4 shows the typical complex impedance plots for these two samples. Only one semicircle appears in the complex impedance spectra. Since the LSMO layer and alloy substrate are electric conductors, the semicircle corresponds to the oxide layer formed at the interface. Figure 5 shows an equivalent circuit used for fitting the impedance spectra of the specimens,

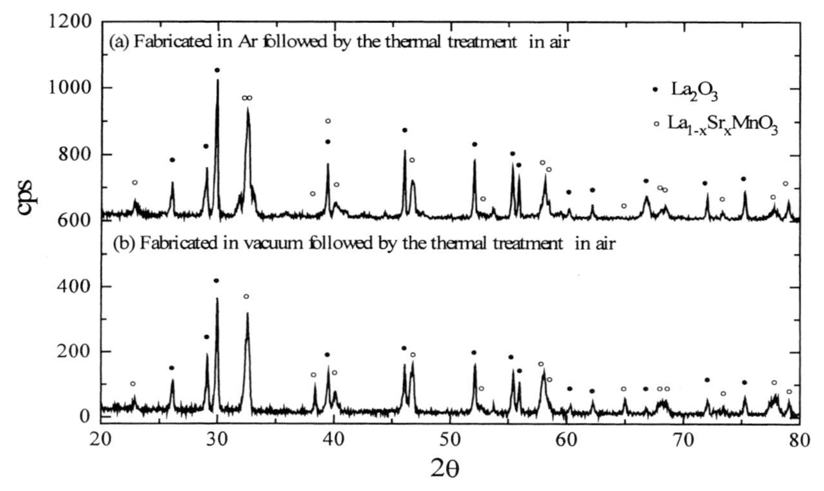

Figure 3. X-ray diffraction analysis of LSMO in the LSMO/fecralloy sample fabricated (a) in flowing Ar and (b) in vacuum and a subsequent thermal treatment at 1200°C in air for 5 h.

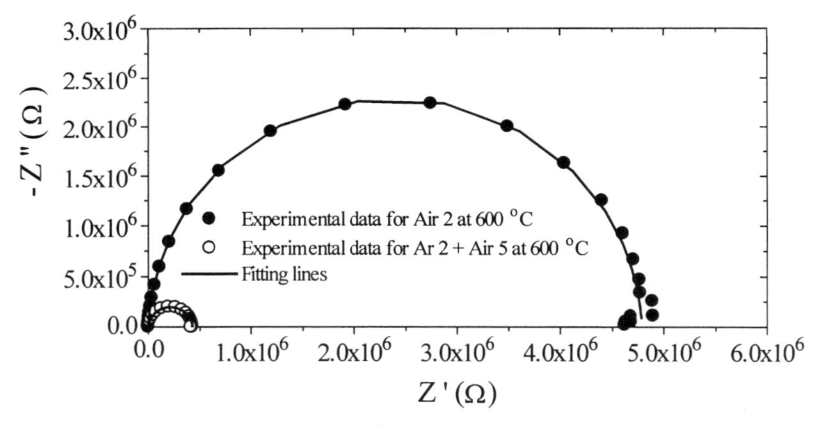

Figure 4. Complex impedance diagrams of the LSMO/fecralloy samples both fabricated in air and fabricated in flowing Ar with a subsequent thermal treatment at 1200°C in air for 5 h.

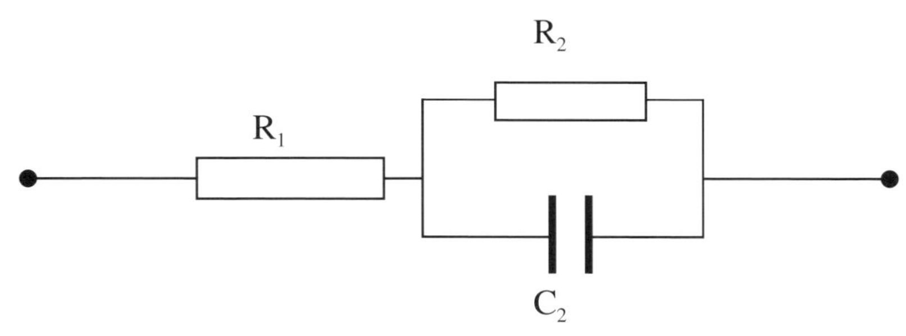

Figure 5. Equivalent circuit representing the LSMO/fecralloy interface; R_2 and C_2 represent the resistance and capacitance of the oxide layer; R_1 represents the resistance of the LSMO layer.

Table 1. Resistance of LSMO layer (R_1), resistance and capacitance of oxide layer (R_2, C_2) for the sample fabricated in air (Air 2) and in flowing Ar with thermal treatments (Ar 2 + Air 5).

Temperature (°C)	Air 2			Ar 2 + Air 5		
	R_1 (Ωcm^2)	R_2 (Ωcm^2)	C_2 (F)	C_2 (F)	R_1 (Ωcm^2)	R_2 (Ωcm^2)
200	7.5	3.15×10^9	9.20×10^{-9}	1.48×10^{-8}	18.0	3.16×10^8
400	6.5	5.71×10^7	1.12×10^{-8}	1.69×10^{-8}	15.0	6.88×10^6
600	5	4.70×10^6	1.32×10^{-8}	1.90×10^{-8}	9.0	4.28×10^5
800	3	6.00×10^5	1.68×10^{-8}	2.60×10^{-8}	7.0	7.58×10^4

where R_1 is the resistance of the LSMO layer since the resistance of the alloy is negligible and R_2 and C_2 are the resistance and capacitance of the oxide layer at the interface respectively. From the fitting of impedance spectra, the values of R_1, R_2 and C_2 for these two samples were obtained and are presented in Table 1. The interfacial electrical resistances at 200, 400, 600 and 800°C for the sample fabricated in air are 3.15×10^9, 5.71×10^7, 4.70×10^6 and 6.00×10^5 Ωcm^2 respectively. The resistances of the sample fabricated in Ar with the subsequent thermal treatment at those temperatures are 3.16×10^8, 6.88×10^6, 4.28×10^5 and 7.58×10^4 Ωcm^2 respectively. Although the resistivity of the oxide layers at both interfaces is lower than that of the pure polycrystalline alumina,[12] the interface electrical resistance is still too high for the Fecralloy to be used as an interconnector in SOFCs.

3.2. *$(La_{0.8}Sr_{0.2})MnO_3$/Ducrolloy Interface*

Figure 6 shows a SEM microgragh of a cross-section of the interface between $(La_{0.8}Sr_{0.2})MnO_3$ and the Ducrolloy. A thick oxide layer (about 10 μm) was formed at the interface after fabrication. EDX analysis indicated that this oxide layer is a high purity Cr_2O_3 phase (see Fig. 7a). Cr cannot be detected within the LSMO grains (Fig. 7b), but Cr and Mn are enriched in the grey regions of the cross section shown in Fig. 6. The grey regions present near the interface are also within the LSMO layer (Fig. 7c and d). The formation of this grey phase is due to the reaction between LSMO phase and CrO_3 vapour evaporated from the Cr_2O_3 oxide layer at the interface at high temperature. X-ray diffraction of the LSMO/ Ducrolloy specimen after post annealing at 1200°C for 10 h shows that this grey phase is a spinel phase $(Mn,Cr)_3O_4$ (Fig. 8). This spinel phase is present in pores and voids of the LSMO layer and seems to form a network surrounding the LSMO grains (Fig. 6).

Figure 9 shows a typical impedance plot of the as prepared LSMO/ducrolloy interface. There are two depressed superimposed semicircles in the complex impedance spectrum, which corresponds to the Cr_2O_3 layer at interface and the $(Mn, Cr)_3O_4$ network in the LSMO layer. To simulate the electrical behaviour of these oxides, equivalent circuits of two parallel R-CPE (constant phase element) in series are assumed (Fig. 10), where R_1 is the resistance of the LSMO layer, R_2 and R_3 are the resistances of Cr_2O_3 and $(Mn,Cr)_3O_4$ phases, and CPE_2 and CPE_3 are capacitance of Cr_2O_3 and $(Mn,Cr)_3O_4$ as function of frequency. The

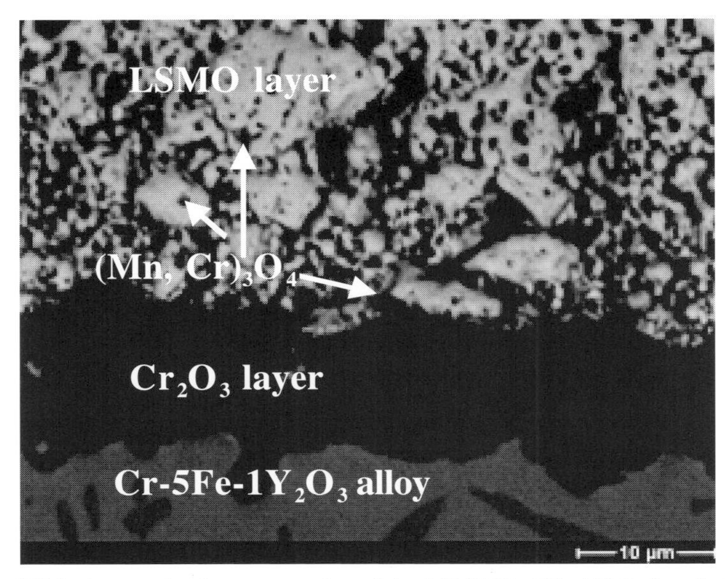

Figure 6. SEM micrograph of a cross-section of the LSMO/Cr–5Fe–1Y_2O_3 interface fabricated at 1200°C for 2 h in air.

constant phase elements (CPEs) are used in the equivalent circuit because the semicircles in the spectra are significantly depressed, indicating the oxide layer is non-homogeneous with multiple relaxation frequencies. Therefore, the CPE should be used here instead of a capacitor to represent the non-homogeneous materials. The impedance of CPE can be described as $Z = 1/[T(i\omega)^P]$, where T and P are two parameters, ω is angular frequency of ac signal and $i = -1^{1/2}$. When $P = 1$, the equation represents the impedance of a capacitor. The deviation of P from 1 corresponds to porosity and non-homogeneity of the materials. For the LSMO/Ducrolloy interface, due to the porous nature and network distribution of $(Mn,Cr)_3O_4$ phase, it is assumed that the parameter P for the simulation of $(Mn, Cr)_3O_4$ phase is smaller than that for the Cr_2O_3 layer. Based on the simulation of impedance spectra in Fig. 9, the parameter P for fitting the first semicircle (at high frequency) is between 0.9 and 1 whereas the P value for fitting the second semicircle (at low frequency) is around 0.7. So, it is likely that the Cr_2O_3 layer corresponds to the first semicircle whereas the $(Mn, Cr)_3O_4$ phase corresponds to the second semicircle, based on the difference in homogeneity between Cr_2O_3 and $(Mn, Cr)_3O_4$. Therefore, the resistor R_2 corresponds to the Cr_2O_3 layer and R_3 corresponds to the $(Mn, Cr)_3O_4$ phase network. The electrical resistances, R_2 and R_3 are high at room temperature ($3.62 \times 10^4\ \Omega\,cm^2$ for the Cr_2O_3 oxide layer and $1.87 \times 10^5\ \Omega\,cm^2$ for the $(Mn, Cr)_3O_4$ network respectively), but become significantly lower at 400°C (1.7 and 7.4 $\Omega\,cm^2$ respectively) and are negligible above 500°C. Fig. 11 shows the R_2 and R_3 as a function of temperature. The resistivity of pure Cr_2O_3 crystal (solid line) and Mn_3O_4 (dash line) obtained from Ref. 13 are also shown in Fig. 11. The ln(R) versus $10^3/T$ for these two oxides from this work do not show linear relationships due to the presence of porosity and impurities in these two phases. The sample after thermal treatment at 1200°C for 10 h

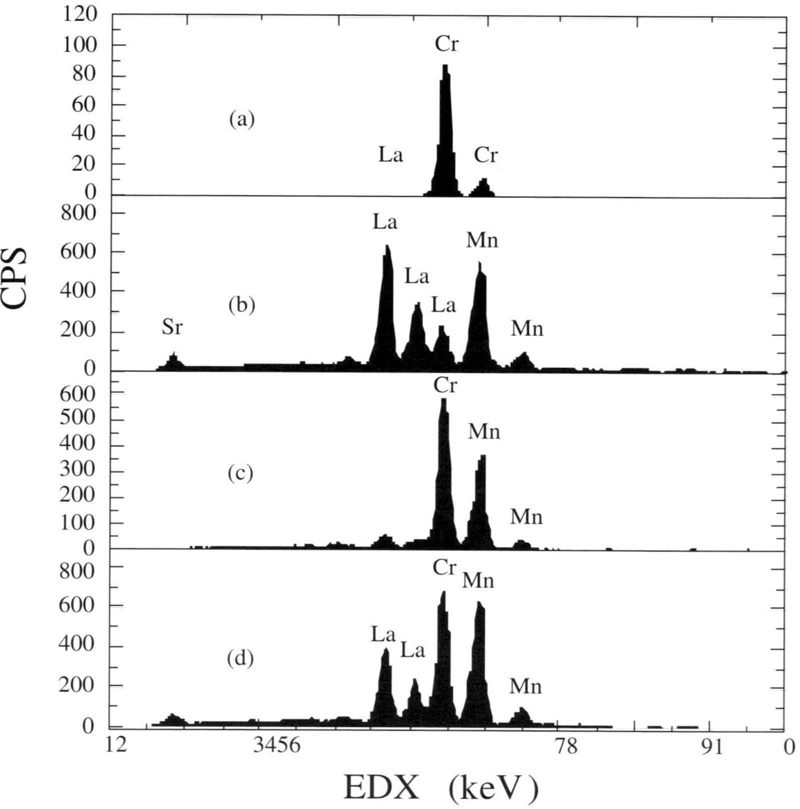

Figure 7. EDX analysis of the LSMO/Ducrolloy specimen; (a) the interfacial layer, (b) LSMO grain (white region in the LSMO region), (c) the grey region in LSMO region close to the interfacial layer and (d) the grey region in LSMO region far from the interfacial layer.

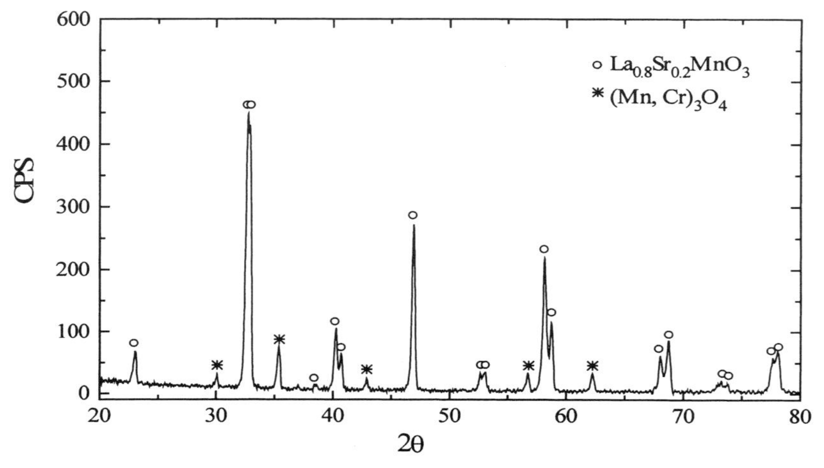

Figure 8. X-ray diffraction analysis of the LSMO phase in the LSMO/Cr–5Fe–1Y_2O_3 sample fabricated at 1200°C in air with subsequent thermal treatment at 1200°C in air for 10 h.

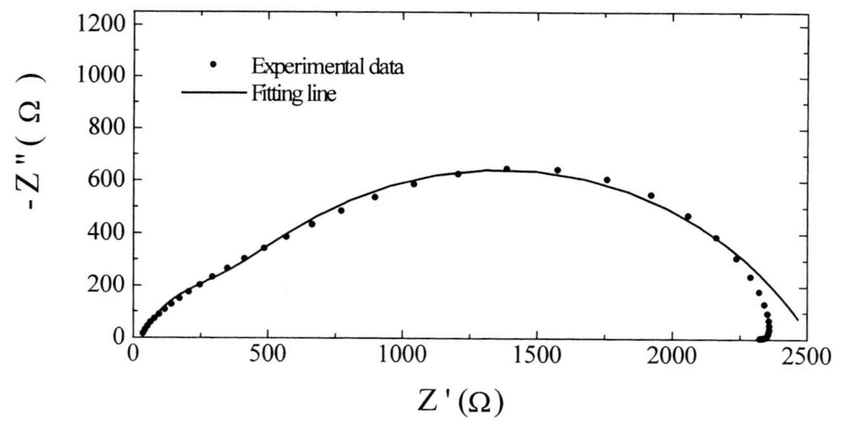

Figure 9. Complex impedance diagram of the LSMO/Cr–5Fe–1Y_2O_3 interface measured at 200°C.

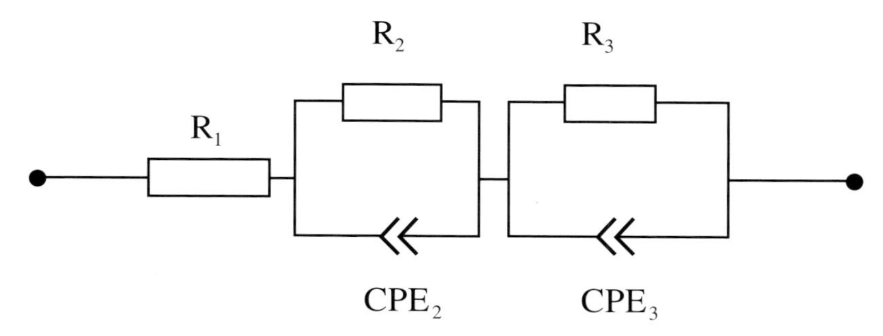

Figure 10. Equivalent circuit representing the LSMO/Cr–5Fe–1Y_2O_3 interface; R_2 and CPE_2 represent the resistance and capacitive effect of the Cr_2O_3 layer; R_3 and CPE_3 represent the resistance and capacitive effect of the $(Mn, Cr)_3O_4$ spinel phase; and R_1 represents the resistance of the LSMO layer.

in air showed an increasing resistance due to the growth of the Cr_2O_3 layer at the interface and the growing $(Mn, Cr)_3O_4$ network in the LSMO layer. However, the resistance of both oxides become significantly lower above 450°C.

Figure 12 shows the resistance of the LSMO layer (R_1 in Fig. 5 and Fig. 10) in the LSMO/fecralloy and LSMO/ducrolloy specimens. In the LSMO/fecralloy specimen, the resistances of the LSMO layers are small for the specimens fabricated both in air and flowing Ar with thermal treatment. In the LSMO/ducrolloy specimen, although the resistance of the LSMO layer was small after fabrication, it increased significantly after thermal treatment in air at 1200°C for 10 h (660 Ω cm^2 at room temperature and 55.0 Ω cm^2 at 500°C). This phenomenon may be caused by the evaporation of CrO_3 into the LSMO layer. This indicates that the main disadvantage of the LSMO/Ducrolloy interface for application in SOFCs is the degradation of LSMO phase due to its reaction with CrO_3 vapour.

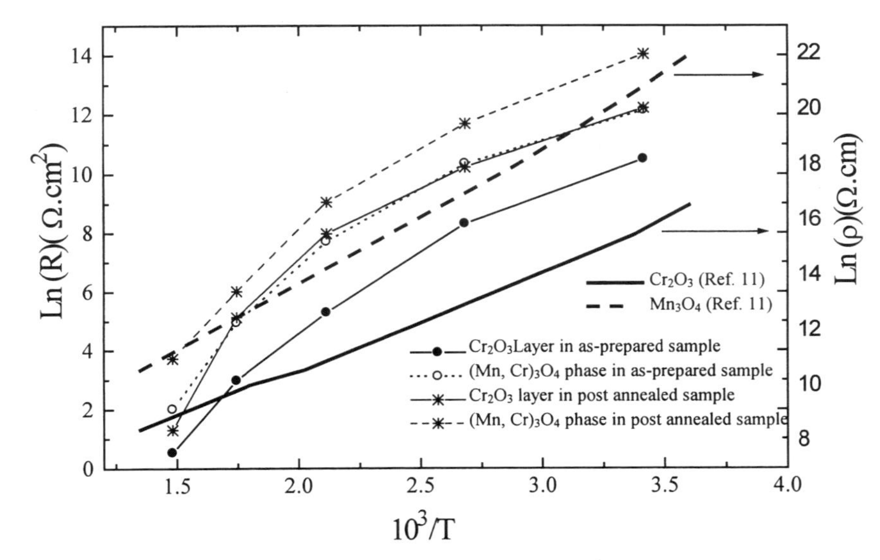

Figure 11. Ln R_2 (resistance of Cr_2O_3 layer), ln R_3 (resistance of (Mn, $Cr)_3O_4$ spinel phase) versus $10^3/T$ for the as prepared and subsequent thermally treated LSMO/Cr–5Fe–1Y_2O_3 interfaces.

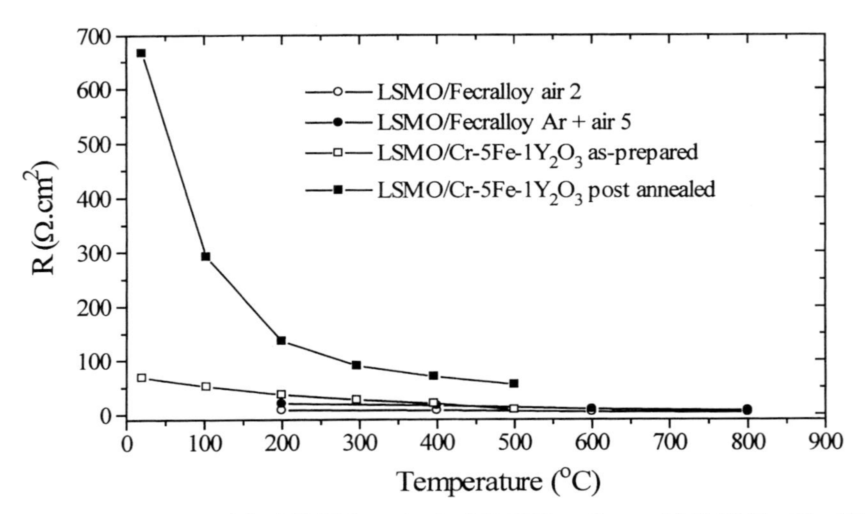

Figure 12. Resistances of the LSMO layer in the LSMO/Fecralloy and LSMO/Cr–5Fe–1Y_2O_3 interfaces as a function of temperature.

4. CONCLUSION

LSMO/fecralloy and LSMO/ducrolloy interfaces have been fabricated and their microstructures examined and related to interface electrical resistances. Experimental results show that the resistance of LSMO/fecralloy interface is very high due to the presence of electrical insulating oxide layer consisting of Al_2O_3 and $M_2O_3.nAl_2O_3$ phases, which were formed during fabrication. The LSMO/Ducrolloy interface shows low electrical resistance although both Cr_2O_3 and $(Mn, Cr)_3O_4$ phases were formed during fabrication. The ducrolloy can be considered as a candidate interconnector material in SOFCs. However, the reaction between CrO_3 vapour and LSMO phase increases the resistivity of LSMO layer, which will affect the stability of the LSMO. Coating of Cr–5Fe–1Y_2O_3 alloy before bonding to LSMO is necessary to prevent the evaporation of CrO_3 phase at the high temperature.

ACKNOWLEDGEMENTS

The authors would like to acknowledge Dr J. Shemilt for useful advice on the preparation of the manuscript. This work is supported by Engineering and Physical Science Research Council (EPSRC, GM/M40271) of United Kingdom.

REFERENCES

1. N. Q. Minh: *J. Am. Ceram. Soc.*, 1993, **76**, 563–588.
2. S. Carter, A. Selcuk, R. J. Chater, J. Kajda, J. A. Kilner and B. C. H. Steele: *Solid State Ionics*, 1992, **53/56**, 597–605.
3. R. E. Williford, T. R. Armstrong and J. D. Gale: *J. Solid State Chem.*, 2000, **149**,(2) 320–326
4. S. Linderoth, P. V. Hendriksen, M. Mogensen and N. Langvad: *J. Mater. Sci.*, 1996, **31**, 5077–5082.
5. S. Srilomsak, D. P. Schilling and H. U. Anderson: *Proc. 1st Int. Symp. on Solid State Fuel Cell*, S. C. Singhal ed., The Electrochemical Society, Pennington, NJ, 1989, 129–140.
6. H. S. Rak, R. S. Dong and M. Dokiya: in *Fuel Cell Seminar. Program and Abstracts*, Courtesy Associates, Washington, DC, 1996, 179–182.
7. W. Thierfelder, H. Greiner and W. Kock: *Proc. 5th Int. Symp. on Solid Oxide Fuel Cells (SOFC-V)*, 1306–15; 1997, The Electrochemical Society, Pennington, NJ, 1997, 1306–1315.
8. P. Kofstad, and R. Bredesen: *Solid State Ionics*, 1992, **52**, 69–75.
9. T. Kadowaki, T. Shiomitsu, E. Matsuda, H. Nakagawa, H. Tsuneizumi and T. Maruyama: *Solid State Ionics*, 1993, **67**, 65–69.
10. W. J. Quadakkers, H. Greiner, M. Hansel, A. Pattanaik, A. S. Khanna and W. Mallener: *Solid State Ionics*, 1996, **91**, 55–67.
11. V. A. Cherepanov, L. Yu-Barkhatova and V. I. Voronin: *J. Solid State Chem.*, 1997, **134**, 38–44.
12. W. D. Kingery, H. K. Bowen and D. R. Uhlmann: *Introduction to Ceramics*, 2nd edn, John Wiley & Son, Inc., New York, NY, 1976, 905.
13. W. D. Kingery, H. K. Bowen and D. R. Uhlmann: *Introduction to Ceramics*, 2nd edn, John Wiley & Son, Inc., New York, NY, 1976, 876.

Author Index

Subject Index